Gene Biotechnology: Concepts and Applications

Gene Biotechnology: Concepts and Applications

Edited by

William Barksdale

Larsen & Keller
www.larsen-keller.com

Gene Biotechnology: Concepts and Applications
Edited by William Barksdale
ISBN: 978-1-63549-651-2 (Hardback)

Larsen & Keller

Published by Larsen and Keller Education,
5 Penn Plaza,
19th Floor,
New York, NY 10001, USA

Cataloging-in-Publication Data

Gene biotechnology : concepts and applications / edited by William Barksdale.
p. cm.
Includes bibliographical references and index.
ISBN 978-1-63549-651-2
1. Genetic engineering. 2. Molecular biology. I. Barksdale, William.
QH442 .G46 2018
660.65--dc23

Printed and bound in China.

For more information regarding Larsen and Keller Education and its products, please visit the publisher's website www.larsen-keller.com

Table of Contents

Preface

As a part of biochemistry, gene biotechnology amalgamates the elements of genetics and biotechnology to understand and experiment with genes in order to develop better genetically modified and cloned organisms. Gene isolation and gene cloning is also a part of gene biotechnology. This book is designed to provide in-depth information about the various concepts, theories, techniques and methods related to gene biotechnology. It presents chapters which deal with the most important topics in this field and provide detailed information about its applications. As this field is emerging at a rapid pace, the contents of this textbook will help the readers understand the modern concepts of the subject.

A detailed account of the significant topics covered in this book is provided below:

Chapter 1- Gene is a constituent of DNA sequence as per which the characteristic of a living organism is determined. Since genes have a particular location in the chromosome, any deference can be found and genetic disease, if any, could be found. Intervention to correct genetic mutation and polymer translation into protein are two important methods in this area. This chapter will provide an integrated understanding of gene therapy.

Chapter 2- Gene therapy can be categorised as somatic and germline gene therapy. In somatic gene therapy, the curing genes affect only the patient and cannot be passed onto the offspring. In germline gene therapy, the germ cells are modified and therefore it becomes heritable. The major categories of gene therapy are dealt with great details in the section.

Chapter 3- In vivo gene therapy involves the introduction of biomolecules directly into the patient's blood stream so that the vector may deliver the therapeutic agent by finding and reaching the target tissue. *In situ* gene therapy involves the direct placement of corrected genes onto the affected area.This chapter discusses the methods of *in situ* and *in vivo* gene therapy in a critical manner providing key analysis to the subject matter.

Chapter 4- The tools used by molecular biologists to introduce genetic material in cells are called viral vectors. Normally a viral vector is created according to particular needs; however, they share a few characteristics. These features include lox toxicity, stability, safe nature, identification and cell type specificity. Lentiviruses are especially used in gene therapy. This chapter has been carefully written to provide an easy understanding of the varied facets of viral vectors.

Chapter 5- The organisms that undergo processes for the modification of their genes are known as transgenic or genetically modified organisms. Mice are the most commonly used genetically modified animals. Easy manipulation of embryos, short gestation periods and availability of complete mouse genome are some of the reasons that mice are used in genetic research. The aspects elucidated in this section are of vital importance, and provide a better understanding of genetically modified organism.

It gives me an immense pleasure to thank our entire team for their efforts. Finally in the end, I would like to thank my family and colleagues who have been a great source of inspiration and support.

Editor

1

Understanding Gene Therapy

Gene is a constituent of DNA sequence as per which the characteristic of a living organism is determined. Since genes have a particular location in the chromosome, any deference can be found and genetic disease, if any, could be found. Intervention to correct genetic mutation and polymer translation into protein are two important methods in this area. This chapter will provide an integrated understanding of gene therapy.

Gene Therapy

Hershey and Chase experiment in 1952 proved that the DNA is a genetic material; the finding revolutionized the field of biology. In 1953, the DNA double helical structure was elucidated by Watson and Crick based on the X-ray diffraction image by Rosalind Franklin. Subsequently the genetic code was elucidated by Nirenberg and Khorana. Further Maniatis and coworkers cloned the globin gene. All these discoveries opened a large arena for manipulating the animal genome and ask the question whether we can treat any heritable disease by using molecular biology tool. The field of gene therapy emerged as a new area for the researchers around the world.

A hereditary unit that consists of sequence of DNA which occupies a specific location on a chromosome and determines a particular characteristic of a living organism is called as Gene . Fluorescence in situ hybridization (FISH) is a method by which a gene is localized with the help of a fluorescent dye in a chromosome. The gene can be visualized for its correct position and location in the chromosome. Any deviation from the normal can be suspected for a potential genetic disease.

Table: Comparison of a eukaryotic cell with a prokaryotic cell:

Cell organelle	Eukaryotic cell	Prokaryotic cell
Nucleus	Present	Absent
Cell type	Usually multicellular	Unicellular
True membrane bound nucleus	Present	Absent
Lysosomes and peroxisomes	Present	Absent
Endoplasmic reticulum and mitochondria	Present	Absent
Ribosomes	Larger	Smaller
Cell wall	Only in plant cells and fungi Chemically simpler	Chemically complex
Plasma membrane with steroids	Yes	Usually no

All the living cells in the universe are divided into either a prokaryotic (primitive nucleus) or a eukaryotic cell (true nucleus). Although they are different from each other in many aspects but both contain genes that code for specific polypeptides required for their perpetuation in the environment.

Gene therapy is a process that involves introduction of genetic material into a person's cell to cure or prevent the disease condition. Researchers around the world are studying gene therapy for a number of diseases, such as cystic fibrosis, hemophilia , cancer , and even influenza, through a number of different approaches.

Basic process of gene therapy involves one of the followings

- Introducing a new specific gene into the body to combat a particular disease.
- Interchanging a disease causing gene with a healthy copy of same gene.
- Inactivating, or "knocking out," of an impaired gene that is not carrying normal function using engineered nucleases such as zinc finger nucleases, engineered I- *CreI* homing endonucleases or the nucleases obtained from TAL effectors. One of the best examples of gene-knockout mediated gene therapy is the knockout of the human CCR5 gene in T-cells to control HIV infection.

The method of gene transfer and its expression is termed as transduction. A gene can be transferred to an individual using a carrier known as a " vector ". Despite the fact that the principle of gene therapy is quite simple but the outcome depends mainly on the vehicles or vectors through which the gene transfer is carried out. Practically gene transfer vectors can be divided either into synthetic or virus based gene transfer systems. With the passage of time many gene transfer vectors have been introduced. Viruses such as adenovirus, adeno-associated virus, and lentiviruses are the most common types of vectors used in gene therapy. The viruses are altered in order to make it safe before using them as a vector for gene therapy. However some risks still exist with gene therapy. The method is still in its inception, but it has been used with varying degree of success.

The first gene therapy experiment was conducted on a 4-years-old girl at National Institute of Health, USA in 1990. She was suffering from a genetic disease having deficiency of adenosine deaminase enzyme required for the proper metabolism of purine. Any malfunctioning of adenosine deaminase leads to severe combined immunodeficiency in affected patients. The white blood cells of the girl was taken out and corrected with a human gene using retrovirus as a vector.

In general, an altered gene can be inserted into a patient using a viral or a non-viral vector. Essentially all cells of the body contain genes, making them suitable for gene therapy. Broadly cells can be divided into somatic (most cells of the body) or germline (eggs or sperm). It is possible to manipulate either somatic cells or germ cells in terms of genetic composition and it also serves as the basis of gene therapy. Gene therapy using germ cells result in permanent changes that are subsequently passed on to next generation. Gene therapy using somatic cells are not passed on to the next generation. Somatic cell gene therapy is safer than the germ line gene therapy because it targets only the affected cells in the patient's body. Somatic cell gene therapy can be done exterior (where cells are modulated *in vitro* and then transplanted back) or interior (where genes are changed *in vivo*).

Gene Doping:

The non-therapeutic use of cells, genes, genetic elements, or of the modulation of gene expression, having the capacity to improve athletic performance is defined as Gene Doping by the World Anti-Doping Agency (WADA).

Qualities of a Suitable Vector

1) An optimal vector should be one that can be made available in a highly intense form using an ideal and reproducible production plan.
2) The vector must be competent of directing the cell type most relevant for the disease, either it be multiplying or non-multiplying cells.
3) Those vectors that can accomplish site-specific integration are more preferred and desirable in case of insertional mutagenesis.
4) A perfect vector should be harmless and should not cause any noxious effect on human health.

Synthetic gene delivery systems rely on direct delivery of genetic message into a specific cell and involve injection of exposed DNA and encapsulation of DNA with liposomes (cationic lipids mixed with nucleic acids), and nanoparticles. Even though this method displays low toxicity and is formed in abundance, gene transfer in common is weak and often temporary. The synthetic gene delivery transfer vectors are also known as non-viral vectors. It is possible that the time to come may witness advanced gene therapy protocols for synthetic gene delivery systems based on the experiences from non viral vector systems.

Viral vector systems are established on replicating viruses with either RNA or DNA genomes that have the potential to deliver genetic message into the host cell. These vectors are either of integrating or of non-integrating type. Integrating viral vectors exhibit constant expression of the defective gene product. Overall, genomes of replicating viruses consist of coding regions and *cis* -acting regulatory elements. The coding sequences include the genetic message of the viral, structural and regulatory proteins and are needed for division of infectious viruses, whereas *cis* -acting sequences are important for wrapping of viral genomes and unification into the host cell. To generate a replication-defective viral vector, the coding domains of the virus are substituted by the genetic message of a therapeutic gene, leaving the *cis* -acting sequences intact.

Monogenic disorders	Cancers
Cystic fibrosis	Gastrointestinal tumors
Hemophilia	Nervous system tumors
Severe combine immuno deficiency	Sarcoma
Familial hypercholesterolemia	Respiratory system tumors

Gene therapy offers promising approaches towards treating many incurable diseases. Many hurdles and challenges are there for a sustained gene therapy approach and scientists around the globe are working to increase the efficacy of gene delivery vectors.

Non-viral Vectors in Gene Therapy

Non viral vectors are biocompatible moieties which can be produced in large scale, easy to manipulate and are easy for cell or tissue targeting. The only disadvantage of non-viral vector is its reduced transfection efficiency in cells or tissues. Non-viral delivery systems utilize different genetic materials such as antisense oligonucleotide (AON), plasmid DNA, small interfering RNA (siRNA), short hairpin RNA (shRNA), and micro RNA (miRNA) that works on electrostatic interaction. On the other hand cationic vectors such as lipid and polymers help in the formation of lipoplexes and polyplexes.

Liposomes in Gene Therapy

The name liposome has been derived from two greek words "lipos" meaning fat and "soma" meaning body. A liposome is a tiny bubble (vehicle) made out of the same material as a cell membrane. Liposomes can be filled with DNA and can be used to deliver for cancer and other diseases. Membranes are usually made of phospholipids which are molecules that have a head group and a tail group. Head is hydrophillic and tail is long hydrocarbon chain and is hydrophobic. Usually phospholipids are bilayer stable membranes. During packaging of the polynucleotide material in liposome, a number of structures are formed. Each structure formed is the most energetically favorable conformation based upon characteristics of specific lipids used. A dependent term known as the structure-packing parameter can be used to suggest what shape the amphiphile will take depending on the size variables.

Lipoplexes and Polyplexes in Gene Therapy

To improve the delivery of DNA into the cell, it must be protected from damage and its entry into the cell must be facilitated. For this lipoplexes and polyplexes are used because of their ability to protect DNA from undesirable degradation during the transfection process. Plasmid DNAs are covered by lipids in an organized structure like a micelle or liposome. When the organized structure is complexed with DNA, it is called lipoplex. There are three types of lipids: cationic, anionic and neutral. Cationic lipids are most preferred for gene delivery because these are positively charged and thus complex well with negatively charged DNA and interacts with the cell membrane, endocytosed by the cell and finally release the DNA.

Naked DNA in Gene Therapy

The simplest non-viral gene delivery system uses naked DNA as a vector. Direct injection of free DNA into certain tissues specially muscles produce high levels of gene expression. It is particularly applied to cancer tissues where the DNA can be injected either directly into the tumor or can be injected into muscle in order to express tumor antigens that might function as a cancer vaccine. It can also be used to treat genetic diseases in the tissues that are available for direct injection such as skin. But the expression is usually non uniform and fails to correct the underlying histological and functional abnormalities of the disease. Though it leads to gene expression but level of expression is much lower than with either viral or liposomal vectors. It is unsuitable for systemic administration due to the presence of serum nuclease. Therefore its application is limited to muscle cells and skin.

Immunotherapy

Immunotherapy comprises of majority of the gene therapy trials for cancer. It has been well established that cancer cells are weakly immunogenic because they are self and by virtue of many ways to down regulate host immune response. Most of the studies related to cancer gene therapy involve manipulation of tumor cells *ex vivo* to enhance the production of interleukins, interferon gamma and tumor necrosis factor. Moreover irradiated tumor cells can be used as a vaccine in many types of cancers. Currently, antigen presenting cells such as dendritic cells isolated from a patient are modulated to enhance the immune response against cancer.

Gene Therapy for Infectious Diseases

Gene therapy may be an attractive tool for the treatment of many important infectious diseases such as HIV. The disease is affecting more than 15 million people worldwide causing decrease in CD4+ cells in the body making patient immunocompromised. The current approaches include targeting siRNA to inhibit virus replication, suicidal gene therapy using adenoviral vectors, and transducing HIV coat protein into patients for inducing immune response. In addition, cloning the immunogenic gene into a suitable plasmid expression vector under the control of tissue specific promoter can be used as a DNA vaccine to elicit immune response against many viral and bacterial diseases.

Virosome

Virosome is a drug or vaccine delivery system. It is a virus-like particle that acts as a vaccine carrier and adjuvant thereby acting as an immune enhancing system. Vaccines that are manufactured using virosome technology show high efficacy with high purity and hence it is a safe and an effective way to vaccinate infants and adults. Virosomes are reconstituted viral envelopes including membrane lipids and viral spike glycoproteins but devoid of viral genome. They are highly effective as vaccine antigens and adjuvants because they stimulate humoral immune response because of the presence of viral glycoproteins. The main advantage of virosomes over other drug delivery system like liposomal and proteoliposomal carrier system is that the virosomes protect pharmaceutically active substances from proteolytic degradation and low pH within endosomes. This helps the contents to be in intact form till it reaches cytoplasm. Influenza virus is most commonly used for virosome production. So, in this case it will have influenza virus' hemagglutinin (HA) and neuraminidase (NA) glycoproteins within the phospholipid bilayer membrane. Various ligands such as cytokines, peptides, monoclonal antibodies (MAbs) can be incorporated into virosome and displayed on the virosomal surface. Due to the presence of influenza HA protein virosomes have unique property of fusion activity. Virosomes bind to the cell surface and enter the cell by receptor mediated endocytosis. Due to the presence of acidic environment in the cell the contents of the virosome are released from the endosome into the cytoplasm of the target cell. Hence virosome technology is a significant technique for the delivery of virus antigens or DNA/RNA encoding specific immune stimulatory molecules. It is target specific and can stimulate both cellular and antibody immune response to maximize protection against the targeted disease. In addition to this, they are completely biodegradable.

Vectors in Gene Therapy

Gene therapy utilizes the delivery of DNA into cells, which can be accomplished by several methods, summarized below. The two major classes of methods are those that use recombinant viruses (sometimes called biological nanoparticles or viral vectors) and those that use naked DNA or DNA complexes (non-viral methods).

Viruses

All viruses bind to their hosts and introduce their genetic material into the host cell as part of their replication cycle. This genetic material contains basic 'instructions' of how to produce more copies of these viruses, hacking the body's normal production machinery to serve the needs of the virus. The host cell will carry out these instructions and produce additional copies of the virus, leading to more and more cells becoming infected. Some types of viruses insert their genome into the host's cytoplasm, but do not actually enter the cell. Others penetrate the cell membrane disguised as protein molecules and enter the cell.

There are two main types of virus infection: lytic and lysogenic. Shortly after inserting its DNA, viruses of the lytic cycle quickly produce more viruses, burst from the cell and infect more cells. Lysogenic viruses integrate their DNA into the DNA of the host cell and may live in the body for many years before responding to a trigger. The virus reproduces as the cell does and does not inflict bodily harm until it is triggered. The trigger releases the DNA from that of the host and employs it to create new viruses.

Retroviruses

The genetic material in retroviruses is in the form of RNA molecules, while the genetic material of their hosts is in the form of DNA. When a retrovirus infects a host cell, it will introduce its RNA together with some enzymes, namely reverse transcriptase and integrase, into the cell. This RNA molecule from the retrovirus must produce a DNA copy from its RNA molecule before it can be integrated into the genetic material of the host cell. The process of producing a DNA copy from an RNA molecule is termed reverse transcription. It is carried out by one of the enzymes carried in the virus, called reverse transcriptase. After this DNA copy is produced and is free in the nucleus of the host cell, it must be incorporated into the genome of the host cell. That is, it must be inserted into the large DNA molecules in the cell (the chromosomes). This process is done by another enzyme carried in the virus called integrase.

Now that the genetic material of the virus has been inserted, it can be said that the host cell has been modified to contain new genes. If this host cell divides later, its descendants will all contain the new genes. Sometimes the genes of the retrovirus do not express their information immediately.

One of the problems of gene therapy using retroviruses is that the integrase enzyme can insert the genetic material of the virus into any arbitrary position in the genome of the host; it randomly inserts the genetic material into a chromosome. If genetic material happens to be inserted in the middle of one of the original genes of the host cell, this gene will be disrupted (insertional mutagenesis). If the gene happens to be one regulating cell division, uncontrolled cell division (i.e., cancer) can occur. This problem has recently begun to be addressed by utilizing zinc finger nucleases or by including certain sequences such as the beta-globin locus control region to direct the site of integration to specific chromosomal sites.

Gene therapy trials using retroviral vectors to treat X-linked severe combined immunodeficiency (X-SCID) represent the most successful application of gene therapy to date. More than twenty patients have been treated in France and Britain, with a high rate of immune system reconstitution observed. Similar trials were restricted or halted in the USA when leukemia was reported in patients treated in the French X-SCID gene therapy trial. To date, four children in the French trial and one in the British trial have developed leukemia as a result of insertional mutagenesis by the retroviral vector. All but one of these children responded well to conventional anti-leukemia treatment. Gene therapy trials to treat SCID due to deficiency of the Adenosine Deaminase (ADA) enzyme (one form of SCID) continue with relative success in the USA, Britain, Ireland, Italy and Japan.

Adenoviruses

Adenoviruses are viruses that carry their genetic material in the form of double-stranded DNA. They cause respiratory, intestinal, and eye infections in humans (especially the common cold). When these viruses infect a host cell, they introduce their DNA molecule into the host. The genetic material of the adenoviruses is not incorporated (transient) into the host cell's genetic material. The DNA molecule is left free in the nucleus of the host cell, and the instructions in this extra DNA molecule are transcribed just like any other gene. The only difference is that these extra genes are not replicated when the cell is about to undergo cell division so the descendants of that cell will not have the extra gene. As a result, treatment with the adenovirus will require readministration in a growing cell population although the absence of integration into the host cell's genome should prevent the type of cancer seen in the SCID trials. This vector system has been promoted for treating cancer and indeed the first gene therapy product to be licensed to treat cancer, Gendicine, is an adenovirus. Gendicine, an adenoviral p53-based gene therapy was approved by the Chinese food and drug regulators in 2003 for treatment of head and neck cancer. Advexin, a similar gene therapy approach from Introgen, was turned down by the US Food and Drug Administration (FDA) in 2008.

Concerns about the safety of adenovirus vectors were raised after the 1999 death of Jesse Gelsinger while participating in a gene therapy trial. Since then, work using adenovirus vectors has focused on genetically crippled versions of the virus.

Envelope Protein Pseudotyping of Viral Vectors

The viral vectors described above have natural host cell populations that they infect most efficiently. Retroviruses have limited natural host cell ranges, and although adenovirus and adeno-associated virus are able to infect a relatively broader range of cells efficiently, some cell types are refractory to infection by these viruses as well. Attachment to and entry into a susceptible cell is mediated by the protein envelope on the surface of a virus. Retroviruses and adeno-associated viruses have a single protein coating their membrane, while adenoviruses are coated with both an envelope protein and fibers that extend away from the surface of the virus. The envelope proteins on each of these viruses bind to cell-surface molecules such as heparin sulfate, which localizes them upon the surface of the potential host, as well as with the specific protein receptor that either induces entry-promoting structural changes in the viral protein, or localizes the virus in endosomes wherein acidification of the lumen induces this refolding of the viral coat. In either case, entry into potential host cells requires a favorable interaction between a protein on the surface of the virus and a protein on the surface of the cell. For the purposes of gene therapy, one might either want to limit or expand the range of cells susceptible to transduction by a gene therapy vector. To this end, many vectors have been developed in which the endogenous viral envelope proteins have been replaced by either envelope proteins from other viruses, or by chimeric proteins. Such chimera would consist of those parts of the viral protein necessary for incorporation into the virion as well as sequences meant to interact with specific host cell proteins. Viruses in which the envelope proteins have been replaced as described are referred to as pseudotyped viruses. For example, the most popular retroviral vector for use in gene therapy trials has been the lentivirus Simian immunodeficiency virus coated with the envelope proteins, G-protein, from Vesicular stomatitis virus. This vector is referred to as VSV G-pseudotyped lentivirus, and infects an almost universal set of cells. This tropism is characteristic of the VSV G-protein with which this vector is coated. Many attempts have been made to limit the tropism of viral vectors to one or a few host cell populations. This advance would allow for the systemic administration of a relatively small amount of vector. The potential for off-target cell modification would be limited, and many concerns from the medical community would be alleviated. Most attempts to limit tropism have used chimeric envelope proteins bearing antibody fragments. These vectors show great promise for the development of "magic bullet" gene therapies.

Replication-competent Vectors

A replication-competent vector called ONYX-015 is used in replicating tumor cells. It was found that in the absence of the E1B-55Kd viral protein, adenovirus caused very rapid apoptosis of infected, p53(+) cells, and this results in dramatically reduced virus progeny and no subsequent spread. Apoptosis was mainly the result of the ability of EIA to inactivate p300. In p53(-) cells, deletion of E1B 55kd has no consequence in terms of apoptosis, and viral replication is similar to that of wild-type virus, resulting in massive killing of cells.

A replication-defective vector deletes some essential genes. These deleted genes are still necessary in the body so they are replaced with either a helper virus or a DNA molecule.

Cis and Trans-acting Elements

Replication-defective vectors always contain a "transfer construct". The transfer construct carries the gene to be transduced or "transgene". The transfer construct also carries the sequences which are necessary for the general functioning of the viral genome: packaging sequence, repeats for replication and, when needed, priming of reverse transcription. These are denominated cis-acting elements, because they need to be on the same piece of DNA as the viral genome and the gene of interest. Trans-acting elements are viral elements, which can be encoded on a different DNA molecule. For example, the viral structural proteins can be expressed from a different genetic element than the viral genome.

Herpes Simplex Virus

The Herpes simplex virus is a human neurotropic virus. This is mostly examined for gene transfer in the nervous system. The wild type HSV-1 virus is able to infect neurons and evade the host immune response, but may still become reactivated and produce a lytic cycle of viral replication. Therefore, it is typical to use mutant strains of HSV-1 that are deficient in their ability to replicate. Though the latent virus is not transcriptionally apparent, it does possess neuron specific promoters that can continue to function normally. Antibodies to HSV-1 are common in humans, however complications due to herpes infection are somewhat rare. Caution for rare cases of encephalitis must be taken and this provides some rationale to using HSV-2 as a viral vector as it generally has tropism for neuronal cells innervating the urogenital area of the body and could then spare the host of severe pathology in the brain.

Non-viral Methods

Non-viral methods present certain advantages over viral methods, with simple large scale production and low host immunogenicity being just two. Previously, low levels of transfection and expression of the gene held non-viral methods at a disadvantage; however, recent advances in vector technology have yielded molecules and techniques with transfection efficiencies similar to those of viruses.

Injection of Naked DNA

This is the simplest method of non-viral transfection. Clinical trials carried out of intramuscular injection of a naked DNA plasmid have occurred with some success; however, the expression has been very low in comparison to other methods of transfection. In addition to trials with plasmids, there have been trials with naked PCR product, which have had similar or greater success. Cellular uptake of naked DNA is generally inefficient. Research efforts focusing on improving the efficiency of naked DNA uptake have yielded several novel methods, such as electroporation, sonoporation, and the use of a “gene gun”, which shoots DNA coated gold particles into the cell using high pressure gas.

Physical Methods to Enhance Delivery

Electroporation

Electroporation is a method that uses short pulses of high voltage to carry DNA across the cell membrane. This shock is thought to cause temporary formation of pores in the cell membrane, allowing DNA molecules to pass through. Electroporation is generally efficient and works across a broad range of cell types. However, a high rate of cell death following electroporation has limited its use, including clinical applications.

More recently a newer method of electroporation, termed electron-avalanche transfection, has been used in gene therapy experiments. By using a high-voltage plasma discharge, DNA was efficiently delivered following very short (microsecond) pulses.

Compared to electroporation, the technique resulted in greatly increased efficiency and less cellular damage.

Gene Gun

The use of particle bombardment, or the gene gun, is another physical method of DNA transfection. In this technique, DNA is coated onto gold particles and loaded into a device which generates a force to achieve penetration of the DNA into the cells, leaving the gold behind on a "stopping" disk.

Sonoporation

Sonoporation uses ultrasonic frequencies to deliver DNA into cells. The process of acoustic cavitation is thought to disrupt the cell membrane and allow DNA to move into cells.

Magnetofection

In a method termed magnetofection, DNA is complexed to magnetic particles, and a magnet is placed underneath the tissue culture dish to bring DNA complexes into contact with a cell monolayer.

Hydrodynamic Delivery

Hydrodynamic delivery involves rapid injection of a high volume of a solution into vasculature (such as into the inferior vena cava, bile duct, or tail vein). The solution contains molecules that are to be inserted into cells, such as DNA plasmids or siRNA, and transfer of these molecules into cells is assisted by the elevated hydrostatic pressure caused by the high volume of injected solution.

Chemical Methods to Enhance Delivery

Oligonucleotides

The use of synthetic oligonucleotides in gene therapy is to deactivate the genes involved in the disease process. There are several methods by which this is achieved. One strategy uses antisense specific to the target gene to disrupt the transcription of the faulty gene. Another uses small molecules of RNA called siRNA to signal the cell to cleave specific unique sequences in the mRNA transcript of the faulty gene, disrupting translation of the faulty mRNA, and therefore expression of the gene. A further strategy uses double stranded oligodeoxynucleotides as a decoy for the transcription factors that are required to activate the transcription of the target gene. The transcription factors bind to the decoys instead of the promoter of the faulty gene, which reduces the transcription of the target gene, lowering expression. Additionally, single stranded DNA oligonucleotides have been used to direct a single base change within a mutant gene. The oligonucleotide is designed to anneal with complementarity to the target gene with the exception of a central base, the target base, which serves as the template base for repair. This technique is referred to as oligonucleotide mediated gene repair, targeted gene repair, or targeted nucleotide alteration.

Lipoplexes

To improve the delivery of the new DNA into the cell, the DNA must be protected from damage and (positively charged). Initially, anionic and neutral lipids were used for the construction of lipoplexes for synthetic vectors. However, in spite of the facts that there is little toxicity associated with them, that they are compatible with body fluids and that there was a possibility of adapting them to be tissue specific; they are complicated and time consuming to produce so attention was turned to the cationic versions.

Cationic lipids, due to their positive charge, were first used to condense negatively charged DNA molecules so as to facilitate the encapsulation of DNA into liposomes. Later it was found that the use of cationic lipids significantly enhanced the stability of lipoplexes. Also as a result of their charge, cationic liposomes interact with the cell membrane, endocytosis was widely believed as the major route by which cells uptake lipoplexes. Endosomes are formed as the results of endocytosis, however, if genes can not be released into cytoplasm by breaking the membrane of endosome, they will be sent to lysosomes where all DNA will be destroyed before they could achieve their functions. It was also found that although cationic lipids themselves could condense and encapsulate DNA into liposomes, the transfection efficiency is very low due to the lack of ability in terms of "endosomal escaping". However, when helper lipids (usually electroneutral lipids, such as DOPE) were added to form lipoplexes, much higher transfection efficiency was observed. Later on, it was figured out that certain lipids have the ability to destabilize endosomal membranes so as to facilitate the escape of DNA from endosome, therefore those lipids are called fusogenic lipids. Although cationic liposomes have been widely used as an alternative for gene delivery vectors, a dose dependent toxicity of cationic lipids were also observed which could limit their therapeutic usages.

The most common use of lipoplexes has been in gene transfer into cancer cells, where the supplied genes have activated tumor suppressor control genes in the cell and decrease the activity of oncogenes. Recent studies have shown lipoplexes to be useful in transfecting respiratory epithelial cells.

Polymersomes

Polymersomes are synthetic versions of liposomes (vesicles with a lipid bilayer), made of amphiphilic block copolymers. They can encapsulate either hydrophilic or hydrophobic contents and can be used to deliver cargo such as DNA, proteins, or drugs to cells. Advantages of polymersomes over liposomes include greater stability, mechanical strength, blood circulation time, and storage capacity.

Polyplexes

Complexes of polymers with DNA are called polyplexes. Most polyplexes consist of cationic polymers and their fabrication is based on self-assembly by ionic interactions. One important difference between the methods of action of polyplexes and lipoplexes is that polyplexes cannot directly release their DNA load into the cytoplasm. As a result, co-transfection with endosome-lytic agents such as inactivated adenovirus was required to facilitate nanoparticle escape from the endocytic vesicle made during particle uptake. However, a better understanding of the mechanisms by which DNA can escape from endolysosomal pathway, i.e. proton sponge effect, has triggered new polymer synthesis strategies such as incorporation of protonable residues in polymer backbone and has revitalized research on polycation-based systems.

Due to their low toxicity, high loading capacity, and ease of fabrication, polycationic nanocarriers demonstrate great promise compared to their rivals such as viral vectors which show high immunogenicity and potential carcinogenicity, and lipid-based vectors which cause dose dependence toxicity. Polyethyleneimine and chitosan are among the polymeric carriers that have been extensively studies for development of gene delivery therapeutics. Other polycationic carriers such as poly(beta-amino esters) and polyphosphoramidate are being added to the library of potential gene carriers. In addition to the variety of polymers and copolymers, the ease of controlling the size, shape, surface chemistry of these polymeric nano-carriers gives them an edge in targeting capability and taking advantage of enhanced permeability and retention effect.

Dendrimers

A dendrimer is a highly branched macromolecule with a spherical shape. The surface of the particle may be functionalized in many ways and many of the properties of the resulting construct are determined by its surface.

In particular it is possible to construct a cationic dendrimer, i.e. one with a positive surface charge. When in the presence of genetic material such as DNA or RNA, charge complimentarity leads to a temporary association of the nucleic acid with the cationic dendrimer. On reaching its destination the dendrimer-nucleic acid complex is then taken into the cell via endocytosis.

In recent years the benchmark for transfection agents has been cationic lipids. Limitations of these competing reagents have been reported to include: the lack of ability to transfect some cell types, the lack of robust active targeting capabilities, incompatibility with animal models, and toxicity. Dendrimers offer robust covalent construction and extreme control over molecule structure, and therefore size. Together these give compelling advantages compared to existing approaches.

Producing dendrimers has historically been a slow and expensive process consisting of numerous slow reactions, an obstacle that severely curtailed their commercial development. The Michigan-based company Dendritic Nanotechnologies discovered a method to produce dendrimers using kinetically driven chemistry, a process that not only reduced cost by a magnitude of three, but also cut reaction time from over a month to several days. These new "Priostar" dendrimers can be specifically constructed to carry a DNA or RNA payload that transfects cells at a high efficiency with little or no toxicity.

Inorganic Nanoparticles

Inorganic nanoparticles, such as gold, silica, iron oxide (ex. magnetofection) and calcium phosphates have been shown to be capable of gene delivery. Some of the benefits of inorganic vectors is in their storage stability, low manufacturing cost and often time, low immunogenicity, and resistance to microbial attack. Nanosized materials less than 100 nm have been shown to efficiently trap the DNA or RNA and allows its escape from the endosome without degradation. Inorganics have also been shown to exhibit improved in vitro transfection for attached cell lines due to their increased density and preferential location on the base of the culture dish. Quantum dots have also been used successfully and permits the coupling of gene therapy with a stable fluorescence marker.

Cell-penetrating Peptides

Cell-penetrating peptides (CPPs), also known as peptide transduction domains (PTDs), are short peptides (< 40 amino acids) that efficiently pass through cell membranes while being covalently or non-covalently bound to various molecules, thus facilitating these molecules' entry into cells. Cell entry occurs primarily by endocytosis but other entry mechanisms also exist. Examples of cargo molecules of CPPs include nucleic acids, liposomes, and drugs of low molecular weight.

CPP cargo can be directed into specific cell organelles by incorporating localization sequences into CPP sequences. For example, nuclear localization sequences are commonly used to guide CPP cargo into the nucleus. For guidance into mitochondria, a mitochondrial targeting sequence can be used; this method is used in protofection (a technique that allows for foreign mitochondrial DNA to be inserted into cells' mitochondria).

Hybrid Methods

Due to every method of gene transfer having shortcomings, there have been some hybrid methods developed that combine two or more techniques. Virosomes are one example; they combine liposomes with an inactivated HIV or influenza virus. This has been shown to have more efficient gene transfer in respiratory epithelial cells than either viral or liposomal methods alone. Other methods involve mixing other viral vectors with cationic lipids or hybridising viruses.

Herpes Simplex Virus

Herpes simplex virus 1 and 2 (HSV-1 and HSV-2), also known as human herpesvirus 1 and 2 (HHV-1 and HHV-2), are two members of the herpesvirus family, Herpesviridae, that infect humans. Both HSV-1 (which produces most cold sores) and HSV-2 (which produces most genital herpes) are ubiquitous and contagious. They can be spread when an infected person is producing and shedding the virus.

In simple terms, herpes simplex 1 is most commonly known as a "cold sore," while herpes simplex 2 is the one known by the public as "herpes," or "genital herpes." According to the World Health Organization 67% of the world population under the age of 50 have HSV-1.

Symptoms of herpes simplex virus infection include watery blisters in the skin or mucous membranes of the mouth, lips, nose or genitals. Lesions heal with a scab characteristic of herpetic disease. Sometimes, the viruses cause very mild or atypical symptoms during outbreaks. However, they can also cause more troublesome forms of herpes simplex. As neurotropic and neuroinvasive viruses, HSV-1 and -2 persist in the body by becoming *latent* and hiding from the immune system in the cell bodies of neurons. After the initial or *primary* infection, some infected people experience sporadic episodes of viral *reactivation* or *outbreaks*. In an outbreak, the virus in a nerve cell becomes active and is transported via the neuron's axon to the skin, where virus replication and shedding occur and cause new sores. It is one of the most common sexually transmitted infections.

Transmission

HSV-1 and -2 are transmitted by contact with an infected person who has reactivations of the virus. Herpes simplex virus (HSV)-2 is periodically shed in the human genital tract, most often asymptomatically, and most sexual transmissions occur during asymptomatic shedding. Asymptomatic reactivation means that the virus causes atypical, subtle or hard to notice symptoms that are not identified as an active herpes infection. In one study, daily genital swab samples found HSV-2 at a median of 12–28% of days among those who have had an outbreak, and 10% of days among those suffering from asymptomatic infection, with many of these episodes occurring without visible outbreak ("subclinical shedding").

In another study, 73 subjects were randomized to receive valaciclovir 1 g daily or placebo for 60 days each in a 2-way crossover design. A daily swab of the genital area was self-collected for HSV-2 detection by polymerase chain reaction, in order to compare the effect of valaciclovir 1 g once daily for 60 days versus placebo on asymptomatic viral shedding in immunocompetent, HSV-2 seropositive subjects without a history of symptomatic genital herpes infection. The study found that valaciclovir significantly reduced shedding during subclinical days compared to placebo, showing a 71% reduction. 84% of subjects had no shedding while receiving valaciclovir versus 54% of subjects on placebo. 88% of patients treated with valaciclovir had no recognized signs or symptoms versus 77% for placebo.

For HSV-2, subclinical shedding may account for most of the transmission. Studies on discordant partners (one infected with HSV-2, one not) show that the transmission rate is approximately 5 per 10,000 sexual contacts. (Effect of Condoms on Reducing the Transmission of Herpes Simplex Virus Type 2 From Men to Women. A Wald, AGM Langenberg, K Link, et al. JAMA. 2001;285(24):3197) Atypical symptoms are often attributed to other causes such as a yeast infection. HSV-1 is often acquired orally during childhood. It may also be sexually transmitted, including contact with saliva, such as kissing and mouth-to-genital contact (oral sex). HSV-2 is primarily a sexually transmitted infection, but rates of HSV-1 genital infections are increasing.

Both viruses may also be transmitted vertically during childbirth, although the real risk is very low. The risk of infection is minimal if the mother has no symptoms or exposed blisters during delivery. The risk is considerable when the mother is infected with the virus for the first time during late pregnancy.

Herpes simplex viruses can affect areas of skin exposed to contact with an infected person (although shaking hands with an infected person does not transmit this disease) . An example of this is herpetic whitlow which is a herpes infection on the fingers. This was a common affliction of dental surgeons prior to the routine use of gloves when conducting treatment on patients.

Virology

Viral Structure

Animal herpes viruses all share some common properties. The structure of herpes viruses consists of a relatively large double-stranded, linear DNA genome encased within an icosahedral protein cage called the capsid, which is wrapped in a lipid bilayer called the envelope. The envelope is joined to the capsid by means of a tegument. This complete particle is known as the virion. HSV-1 and HSV-2 each contain at least 74 genes (or open reading frames, ORFs) within their genomes, although speculation over gene crowding allows as many as 84 unique protein coding genes by 94 putative ORFs. These genes encode a variety of proteins involved in forming the capsid, tegument and envelope of the virus, as well as controlling the replication and infectivity of the virus.

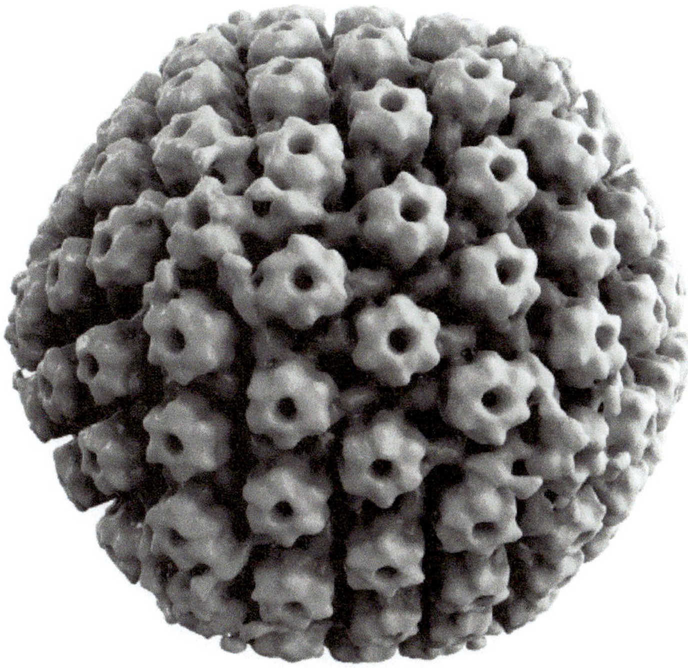

3D reconstruction of the Herpes simplex virus type 1 (HSV-1) capsid

The genomes of HSV-1 and HSV-2 are complex and contain two unique regions called the long unique region (U_L) and the short unique region (U_S). Of the 74 known ORFs, U_L contains 56 viral genes, whereas U_S contains only 12. Transcription of HSV genes is catalyzed by RNA polymerase II of the infected host. Immediate early genes, which encode proteins that regulate the expression of *early* and *late* viral genes, are the first to be expressed following infection. Early gene expression follows, to allow the synthesis of enzymes involved in DNA replication and the production of certain envelope glycoproteins. Expression of late genes occurs last; this group of genes predominantly encode proteins that form the virion particle.

Five proteins from (U_L) form the viral capsid; UL6, UL18, UL35, UL38 and the major capsid protein UL19.

Cellular Entry

Entry of HSV into a host cell involves several glycoproteins on the surface of the enveloped virus binding to their transmembrane receptors on the cell surface. Many of these receptors are then pulled inwards by the cell, which is thought to open a ring of three gHgL heterodimers stabilizing a compact conformation of the gB glycoprotein, so that it springs out and punctures the cell membrane. The envelope covering the virus particle then fuses with the cell membrane, creating a pore through which the contents of the viral envelope enters the host cell.

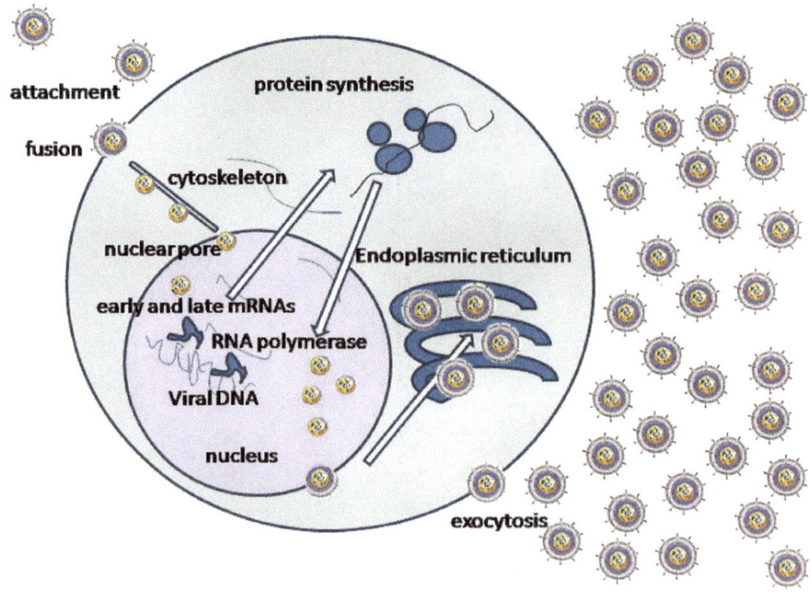

A simplified diagram of HSV replication

The sequential stages of HSV entry are analogous to those of other viruses. At first, complementary receptors on the virus and the cell surface bring the viral and cell membranes into proximity. Interactions of these molecules then form a stable entry pore through which the viral envelope contents are introduced to the host cell. The virus can also be endocytosed after binding to the receptors, and the fusion could occur at the endosome. In electron micrographs the outer leaflets of the viral and cellular lipid bilayers have been seen merged; this *hemifusion* may be on the usual path to entry or it may usually be an arrested state more likely to be captured than a transient entry mechanism.

In the case of a herpes virus, initial interactions occur when two viral envelope glycoprotein called glycoprotein C (gC) and glycoprotein B (gB) bind to a cell surface particle called heparan sulfate. Next, the major receptor binding protein, glycoprotein D (gD), binds specifically to at least one of three known entry receptors. These cell receptors include herpesvirus entry mediator (HVEM), nectin-1 and 3-O sulfated heparan sulfate. The nectin receptors usually produce cell-cell adhesion, to provide a strong point of attachment for the virus to the host cell. These interactions bring the membrane surfaces into mutual proximity and allow for other glycoproteins embedded in the viral envelope to interact with other cell surface molecules. Once bound to the HVEM, gD changes its conformation and interacts with viral glycoproteins H (gH) and L (gL), which form a complex. The interaction of these membrane proteins may result in a hemifusion state. gB interaction with the gH/gL complex creates an entry pore for the viral capsid. gB interacts with glycosaminoglycans on the surface of the host cell.

Genetic Inoculation

After the viral capsid enters the cellular cytoplasm, it is transported to the cell nucleus. Once attached to the nucleus at a *nuclear entry pore*, the capsid ejects its DNA contents via the capsid *portal*. The capsid portal is formed by twelve copies of portal protein, UL6, arranged as a ring; the proteins contain a leucine zipper sequence of amino acids which allow them to adhere to each other. Each icosahedral capsid contains a single portal, located in one vertex. The DNA exits the capsid in a single linear segment.

Immune Evasion

HSV evades the immune system through interference with MHC class I antigen presentation on the cell surface, by blocking TAP or the transporter associated with antigen processing induced by the secretion of ICP-47 by HSV. In the host cell, TAP transports digested viral antigen epitope peptides from the cytosol to the endoplasmic reticulum, allowing these epitopes to be combined with MHC class I molecules and presented on the surface of the cell. Viral epitope presentation with MHC class I is a requirement for activation of cytotoxic T-lymphocytes (CTLs), the major effectors of the cell-mediated immune response against virally-infected cells. ICP-47 prevents initiation of a CTL-response against HSV, allowing the virus to survive for a protracted period in the host.

Replication

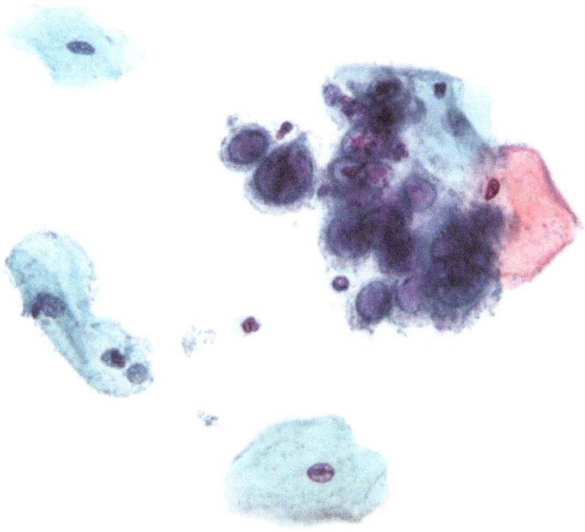

Micrograph showing the viral cytopathic effect of HSV (multi-nucleation, ground glass chromatin).

Following infection of a cell, a cascade of herpes virus proteins, called *immediate-early*, *early*, and *late*, are produced. Research using flow cytometry on another member of the herpes virus family, Kaposi's sarcoma-associated herpesvirus, indicates the possibility of an additional lytic stage, *delayed-late*. These stages of lytic infection, particularly *late lytic*, are distinct from the latency stage. In the case of HSV-1, no protein products are detected during latency, whereas they are detected during the lytic cycle.

The early proteins transcribed are used in the regulation of genetic replication of the virus. On entering the cell, an α-TIF protein joins the viral particle and aids in immediate-early transcription. The virion host shutoff protein (VHS or UL41) is very important to viral replication. This enzyme shuts off protein synthesis in the host, degrades host mRNA, helps in viral replication, and regulates gene expression of viral proteins. The viral genome immediately travels to the nucleus but the VHS protein remains in the cytoplasm.

The late proteins form the capsid and the receptors on the surface of the virus. Packaging of the viral particles — including the genome, core and the capsid - occurs in the nucleus of the cell. Here, concatemers of the viral genome are separated by cleavage and are placed into pre-formed capsids. HSV-1 undergoes a process of primary and secondary envelopment. The primary envelope is acquired by budding into the inner nuclear membrane of the cell. This then fuses with the outer nuclear membrane releasing a naked capsid into the cytoplasm. The virus acquires its final envelope by budding into cytoplasmic vesicles.

Latent Infection

HSVs may persist in a quiescent but persistent form known as *latent infection*, notably in neural ganglia. HSV-1 tends to reside in the trigeminal ganglia, while HSV-2 tends to reside in the sacral ganglia, but these are tendencies only, not fixed behavior. During latent infection of a cell, HSVs express latency associated transcript (LAT) RNA. LAT regulates the host cell genome and interferes with natural cell death mechanisms. By maintaining the host cells, LAT expression preserves a reservoir of the virus, which allows subsequent, usually symptomatic, periodic recurrences or "outbreaks" characteristic of non-latency. Whether or not recurrences are symptomatic, viral shedding occurs to infect a new host. A protein found in neurons may bind to herpes virus DNA and regulate latency. Herpes virus DNA contains a gene for a protein called ICP4, which is an important transactivator of genes associated with lytic infection in HSV-1. Elements surrounding the gene for ICP4 bind a protein known as the human neuronal protein Neuronal Restrictive Silencing Factor (NRSF) or human Repressor Element Silencing Transcription Factor (REST). When bound to the viral DNA elements, histone deacetylation occurs atop the ICP4 gene sequence to prevent initiation of transcription from this gene, thereby preventing transcription of other viral genes involved in the lytic cycle. Another HSV protein reverses the inhibition of ICP4 protein synthesis. ICP0 dissociates NRSF from the *ICP4* gene and thus prevents silencing of the viral DNA.

Germ Cell

A germ cell is any biological cell that gives rise to the gametes of an organism that reproduces sexually. In many animals, the germ cells originate in the primitive streak and migrate via the gut of an embryo to the developing gonads. There, they undergo meiosis, followed by cellular differentiation into mature gametes, either eggs or sperm. Unlike animals, plants do not have germ cells designated in early development. Instead, germ cells can arise from somatic cells in the adult (such as the floral meristem of flowering plants).

Introduction

Multicellular eukaryotes are made of two fundamental cell types. Germ cells produce gametes and are the only cells that can undergo meiosis as well as mitosis. These cells are sometimes said to be immortal because they are the link between generations. Somatic cells are all the other cells that form the building blocks of the body and they only divide by mitosis. The lineage of germ cells is called germ line. Germ cell specification begins during cleavage in many animals or in the epiblast during gastrulation in birds and mammals. After transport, involving passive movements and active migration, germ cells arrive at the developing gonads. In humans, sexual differentiation starts approximately 6 weeks after conception. The end-products of the germ cell cycle are the egg or sperm.

Under special conditions *in vitro* germ cells can acquire properties similar to those of embryonic stem cells (ES). The underlying mechanism of that change is still unknown. These changed cells are then called embryonic germ cells (EG). Both EG and ES are pluripotent in vitro, but only ES has proven pluripotency in vivo. Recent studies have demonstrated that it is possible to give rise to primordial germ cells from ES.

Specification

There are two mechanisms to establish the germ cell lineage in the embryo. The first way is called preformistic and involves that the cells destined to become germ cells inherit the specific germ cell determinants present in the germ plasm (specific area of the cytoplasm) of the egg (ovum). The unfertilized egg of most animals is asymmetrical: different regions of the cytoplasm contain different amounts of mRNA and proteins. By this, germ cells obtained by the first divisions of the fertilized egg are characterized by specific molecules of a particular region of the egg cytoplasm.

The second way is found in birds and mammals, where germ cells are not specified by such determinants but by signals controlled by zygotic genes. In mammals, a few cells of the early embryo are induced by signals of neighboring cells to become primordial germ cells. Mammalian eggs are somewhat symmetrical and after the first divisions of the fertilized egg, the produced cells are all totipotent. This means that they can differentiate in any cell type in the body and thus germ cells. Specification of primordial germ cells in the laboratory mouse is initiated by high levels of bone morphogenetic protein (BMP) signaling, which activates expression of the transcription factors Blimp-1/Prdm1 and Prdm14.

Migration

Primordial germ cells, germ cells that still have to reach the gonads, also known as PGCs, precursor germ cells or gonocytes, divide repeatedly on their migratory route through the gut and into the developing gonads.

Invertebrates

In the model organism *Drosophila*, pole cells passively move from the posterior end of the embryo to the posterior midgut because of the infolding of the blastoderm. Then they actively move through the gut into the mesoderm. Endodermal cells differentiate and together with Wunen proteins they induce the migration through the gut. Wunen proteins are chemorepellents that lead the germ cells away from the endoderm and into the mesoderm. After splitting into two populations, the germ cells continue migrating laterally and in parallel until they reach the gonads. Columbus proteins, chemoattractants, stimulate the migration in the gonadal mesoderm.

Vertebrates

In the *Xenopus* egg, the germ cell determinants are found in the most vegetal blastomeres. These presumptive PGCs are brought to the endoderm of the blastocoel by gastrulation. They are determined as germ cells when gastrulation is completed. Migration from the hindgut along the gut and across the dorsal mesentery then takes place. The germ cells split into two populations and move to the paired gonadal ridges. Migration starts with 3-4 cells that undergo three rounds of cell division so that about 30 PGCs arrive at the gonads. On the migratory path of the PGCs, the orientation of underlying cells and their secreted molecules such as fibronectin play an important role.

Mammals have a migratory path comparable to that in *Xenopus*. Migration begins with 50 gonocytes and about 5,000 PGCs arrive at the gonads. Proliferation occurs also during migration and lasts for 3–4 weeks in humans.

PGCs come from the epiblast and migrate subsequently into the mesoderm, the endoderm and the posterior of the yolk sac. Migration then takes place from the hindgut along the gut and across the dorsal mesentery to reach the gonads (4.5 weeks in human beings). Fibronectin maps here also a polarized network together with other molecules. The somatic cells on the path of germ cells provide them attractive, repulsive, and survival signals. But germ cells also send signals to each other.

In reptiles and birds, germ cells use another path. PGCs come from the epiblast and move to the hypoblast to form the germinal crescent (anterior extraembryonic structure). The gonocytes then squeeze into blood vessels and use the circulatory system for transport. They squeeze out of the vessels when they are at height of the gonadal ridges. Cell adhesion on the endothelium of the blood vessels and molecules such as chemoattractants are probably involved in helping PGCs migrate.

The Sry Gene of the Y Chromosome

The *SRY* (Sex-determining Region of the Y chromosome) directs male development in mammals by inducing the somatic cells of the gonadal ridge to develop into a testis, rather than an ovary. *Sry* is expressed in a small group of somatic cells of the gonads and influences these cells to become Sertoli cells (supporting cells in testis). Sertoli cells are responsible for sexual development along a male pathway in many ways. One of these ways involves stimulation of the arriving primordial cells to differentiate into sperm. In the absence of the *Sry* gene, primordial germ cells differentiate into eggs. Removing genital ridges before they start to develop into testes or ovaries results in the development of a female, independent of the carried sex chromosome.

Gametogenesis

Gametogenesis, the development of diploid germ cells into either haploid eggs or sperm, (respectively oogenesis and spermatogenesis) is different for each species but the general stages are similar. Oogenesis and spermatogenesis have many features in common, they both involve:

- Meiosis
- Extensive morphological differentiation
- Incapacity of surviving for very long if fertilization does not occur

Despite their homologies they also have major differences:

- Spermatogenesis has equivalent meiotic divisions resulting in four equivalent spermatids while oogenic meiosis is asymmetrical: only one egg is formed together with three polar bodies.
- Different timing of maturation: oogenic meiosis is interrupted at one or more stages (for a long time) while spermatogenic meiosis is rapid and uninterrupted.

Oogenesis

After migration primordial germ cells will become oogonia in the forming gonad (ovary). The oogonia proliferate extensively by mitotic divisions, up to 5-7 million cells in humans. But then many of these oogonia die and about 50,000 remain. These cells differentiate into primary oocytes. In week 11-12 *post coitus* the first meiotic division begins (before birth for most mammals) and remains arrested in prophase I from a few days to many years depending on the species. It is in this period or in some cases at the beginning of sexual maturity that the primary oocytes secrete proteins to form a coat called zona pellucida and they also produce cortical granules containing enzymes and proteins needed for fertilization. Meiosis stands by because of the follicular granulosa cells that send inhibitory signals through gap junctions and the zona pellucida. Sexual maturation is the beginning of periodic ovulation. Ovulation is the regular release of one oocyte from the ovary into the reproductive tract and is preceded by follicular growth. A few follicle cells are stimulated to grow but only one oocyte is ovulated. A primordial follicle consists of an epithelial layer of follicular granulosa cells enclosing an oocyte. The pituitary gland secrete follicle-stimulating hormones (FSHs) that stimulate follicular growth and oocyte maturation. The thecal cells around each follicle secrete estrogen. This hormone stimulates the production of FSH receptors on the follicular granulosa cells and has at the same time a negative feedback on FSH secretion. This results in a competition between the follicles and only the follicle with the most FSH receptors survives and is ovulated. Meiotic division I goes on in the ovulated oocyte stimulated by luteinizing hormones (LHs) produced by the pituitary gland. FSH and LH block the gap junctions between follicle cells and the oocyte therefore inhibiting communication between them. Most follicular granulosa cells stay around the oocyte and so form the cumulus layer. Large non-mammalian oocytes accumulate egg yolk, glycogen, lipids, ribosomes, and the mRNA needed for protein synthesis during early embryonic growth. These intensive RNA biosynthese are mirrored in the structure of the chromosomes, which decondense and form lateral loops giving them a lampbrush appearance. Oocyte maturation is the following phase of oocyte development. It occurs at sexual maturity when hormones stimulate the oocyte to complete meiotic division I. The meiotic division I produces 2 cells differing in size: a small polar body and a large secondary oocyte. The secondary oocyte undergoes meiotic division II and that results in the formation of a second small polar body and a large mature egg, both being haploid cells. The polar bodies degenerate. Oocyte maturation stands by at metaphase II in most vertebrates. During ovulation, the arrested secondary oocyte leaves the ovary and matures rapidly into an egg ready for fertilization. Fertilization will cause the egg to complete meiosis II. In human females there is proliferation of the oogonia in the fetus, meiosis starts then before birth and stands by at meiotic division I up to 50 years, ovulation begins at puberty.

Egg Growth

A 10 - 20 µm large somatic cell generally needs 24 hours to double its mass for mitosis. By this way it would take a very long time for that cell to reach the size of a mammalian egg with a diameter of 100 µm (some insects have eggs of about 1,000 µm or greater). Eggs have therefore special mechanisms to grow to their large size. One of these mechanisms is to have extra copies of genes: meiotic division I is paused so that the oocyte grows while it contains two diploid chromosome sets. Some species produce many extra copies of genes, such as amphibians, which may have up to 1 or 2 million copies. A complementary mechanism is partly dependent on syntheses of other cells. In amphibians, birds, and insects, yolk is made by the liver (or its equivalent) and secreted into the blood. Neighboring accessory cells in the ovary can also provide nutritive help of two types. In some invertebrates some oogonia become nurse cells. These cells are connected by cytoplasmic bridges with oocytes. The nurse cells of insects provide oocytes macromolecules such as proteins and mRNA. Follicular granulosa cells are the second type of accessory cells in the ovary in both invertebrates and vertebrates. They form a layer around the oocyte and nourish them with small molecules, no macromolecules, but eventually their smaller precursor molecules, by gap junctions.

Spermatogenesis

Mammalian spermatogenesis is representative for most animals. In human males, spermatogenesis begins at puberty in seminiferous tubules in the testicles and go on continuously. Spermatogonia are immature germ cells. They proliferate continuously by mitotic divisions around the outer edge of the seminiferous tubules, next to the basal lamina. Some of these cells stop proliferation and differentiate into primary spermatocytes. After they proceed through the first meiotic division, two secondary spermatocytes are produced. The two secondary spermatocytes undergo the second meiotic division to form four haploid spermatids. These spermatids differentiate morphologically into sperm by nuclear condensation, ejection of the cytoplasm and formation of the acrosome and flagellum.

The developing male germ cells do not complete cytokinesis during spermatogenesis. Consequently, cytoplasmic bridges assure connection between the clones of differentiating daughter cells to form a syncytium. In this way the haploid cells are supplied with all the products of a complete diploid genome. Sperm that carry a Y chromosome, for example, is supplied with essential molecules that are encoded by genes on the X chromosome.

Evolution

The herpes simplex 1 genomes can be classified into six clades. Four of these occur in East Africa, one in East Asia and one in Europe and North America. This suggests that the virus may have originated in East Africa. The most recent common ancestor of the Eurasian strains appears to have evolved ~60,000 years ago. The East Asian HSV-1 isolates have an unusual pattern that is currently best explained by the two waves of migration responsible for the peopling of Japan.

The mutation rate has been estimated to be $\sim 1.38 \times 10^{-7}$ substitutions/site/year. In clinical setting, the mutations in either the thymidine kinase gene or DNA polymerase gene has caused resistance to aciclovir. However, most of the mutations occur in the thymidine kinase gene rather than the DNA polymerase gene.

Treatment

Herpes viruses establish lifelong infections, and the virus cannot yet be eradicated from the body. Treatment usually involves general-purpose antiviral drugs that interfere with viral replication, reduce the physical severity of outbreak-associated lesions, and lower the chance of transmission to others. Studies of vulnerable patient populations have indicated that daily use of antivirals such as aciclovir and valaciclovir can reduce reactivation rates.

Alzheimer's Disease

In the presence of a certain gene variation (APOE-epsilon4 allele carriers), a possible link between HSV-1 (i.e., the virus that causes cold sores or oral herpes) and Alzheimer's disease was reported in 1979. HSV-1 appears to be particularly damaging to the nervous system and increases one's risk of developing Alzheimer's disease. The virus interacts with the components and receptors of lipoproteins, which may lead to the development of Alzheimer's disease. This research identifies HSVs as the pathogen most clearly linked to the establishment of Alzheimer's. According to a study done in 1997, without the presence of the gene allele, HSV-1 does not appear to cause any neurological damage or increase the risk of Alzheimer's. However, a more recent prospective study published in 2008 with a cohort of 591 people showed a statistically significant difference between patients with antibodies indicating recent reactivation of HSV and those without these antibodies in the incidence of Alzheimer's disease, without direct correlation to the APOE-epsilon4 allele. It should be noted that the trial had a small sample of patients who did not have the antibody at baseline, so the results should be viewed as highly uncertain. In 2011 Manchester University scientists showed that treating HSV1-infected cells with antiviral agents decreased the accumulation of β-amyloid and P-tau, and also decreased HSV-1 replication.

Multiplicity Reactivation

Multiplicity reactivation (MR) is the process by which viral genomes containing inactivating damage interact within an infected cell to form a viable viral genome. MR was originally discovered with the bacterial virus bacteriophage T4, but was subsequently also found with pathogenic viruses including influenza virus, HIV-1, adenovirus simian virus 40, vaccinia virus, reovirus, poliovirus and herpes simplex virus.

When HSV particles are exposed to doses of a DNA damaging agent that would be lethal in single infections, but are then allowed to undergo multiple infection (i.e. two or more viruses per host cell), MR is observed. Enhanced survival of HSV-1 due to MR occurs upon exposure to different DNA damaging agents, including methyl methanesulfonate, trimethylpsoralen (which causes inter-strand DNA cross-links), and UV light. After treatment of genetically marked HSV with trimethylpsoralen, recombination between the marked viruses increases, suggesting that trimethylpsoralen damage stimulates recombination. MR of HSV appears to partially depend on the host cell recombinational repair machinery since skin fibroblast cells defective in a component of this machinery (i.e. cells from Bloom's syndrome patients) are deficient in MR. These observations suggest that MR in HSV infections involves genetic recombination between damaged viral genomes resulting in production of viable progeny viruses. HSV-1, upon infecting host cells, induces inflammation and oxidative stress. Thus it appears that the HSV genome may be subjected to oxidative DNA damage during infection, and that MR may enhance viral survival and virulence under these conditions.

Use as an anti-cancer Agent

Herpes simplex virus is considered as a potential therapy for cancer and has been extensively clinically tested to assess its oncolytic (cancer killing) ability. Interim overall survival data from Amgen's phase 3 trial of a genetically-attenuated herpes virus suggests efficacy against melanoma.

Use in Neuronal Connection Tracing

Herpes simplex virus is also used as a transneuronal tracer defining connections among neurons by virtue of traversing synapses.

Other Related Outcomes

Herpes simplex virus is likely the most common cause of Mollaret's meningitis, and, in worse case scenarios, can lead to a potentially fatal case of herpes simplex encephalitis.

Gene Expression

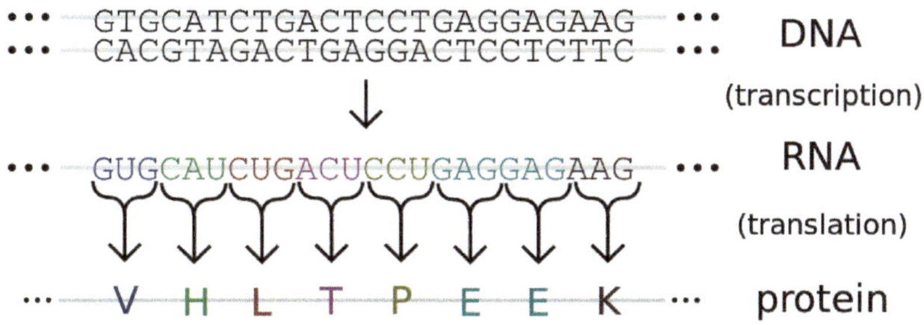

Genes are expressed by being transcribed into RNA, and this transcript may then be translated into protein.

Gene expression is the process by which information from a gene is used in the synthesis of a functional gene product. These products are often proteins, but in non-protein coding genes such as transfer RNA (tRNA) or small nuclear RNA (snRNA) genes, the product is a functional RNA. The process of gene expression is used by all known life—

eukaryotes (including multicellular organisms), prokaryotes (bacteria and archaea), and utilized by viruses—to generate the macromolecular machinery for life.

Several steps in the gene expression process may be modulated, including the transcription, RNA splicing, translation, and post-translational modification of a protein. Gene regulation gives the cell control over structure and function, and is the basis for cellular differentiation, morphogenesis and the versatility and adaptability of any organism. Gene regulation may also serve as a substrate for evolutionary change, since control of the timing, location, and amount of gene expression can have a profound effect on the functions (actions) of the gene in a cell or in a multicellular organism.

In genetics, gene expression is the most fundamental level at which the genotype gives rise to the phenotype, i.e. observable trait. The genetic code stored in DNA is "interpreted" by gene expression, and the properties of the expression give rise to the organism's phenotype. Such phenotypes are often expressed by the synthesis of proteins that control the organism's shape, or that act as enzymes catalysing specific metabolic pathways characterising the organism. Regulation of gene expression is thus critical to an organism's development.

Mechanism

Transcription

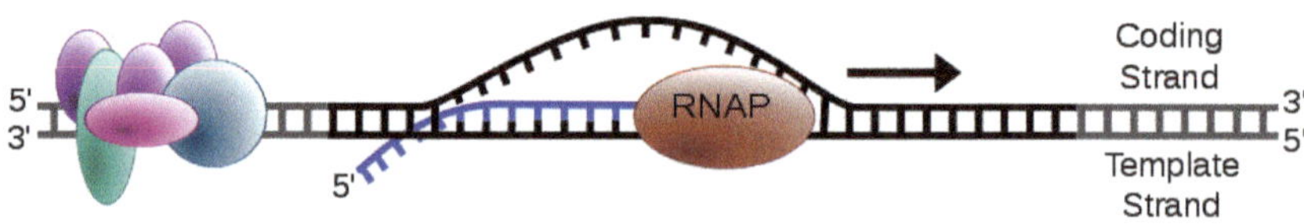

The process of transcription is carried out by RNA polymerase (RNAP), which uses DNA (black) as a template and produces RNA (blue).

A gene is a stretch of DNA that encodes information. Genomic DNA consists of two antiparallel and reverse complementary strands, each having 5' and 3' ends. With respect to a gene, the two strands may be labeled the "template strand," which serves as a blueprint for the production of an RNA transcript, and the "coding strand," which includes the DNA version of the transcript sequence. (Perhaps surprisingly, the "coding strand" is not physically involved in the coding process because it is the "template strand" that is read during transcription.)

The production of the RNA copy of the DNA is called transcription, and is performed in the nucleus by RNA polymerase, which adds one RNA nucleotide at a time to a growing RNA strand as per the complementarity law of the bases. This RNA is complementary to the template 3' → 5' DNA strand, which is itself complementary to the coding 5' → 3' DNA strand. Therefore, the resulting 5' → 3' RNA strand is identical to the coding DNA strand with the exception that thymines (T) are replaced with uracils (U) in the RNA. A coding DNA strand reading "ATG" is indirectly transcribed through the non-coding strand as "AUG" in RNA.

In prokaryotes, transcription is carried out by a single type of RNA polymerase, which needs a DNA sequence called a Pribnow box as well as a sigma factor (σ factor) to start transcription. In eukaryotes, transcription is performed by three types of RNA polymerases, each of which needs a special DNA sequence called the promoter and a set of DNA-binding proteins—transcription factors—to initiate the process. RNA polymerase I is responsible for transcription of ribosomal RNA (rRNA) genes. RNA polymerase II (Pol II) transcribes all protein-coding genes but also some non-coding RNAs (e.g., snRNAs, snoRNAs or long non-coding RNAs). Pol II includes a C-terminal domain (CTD) that is rich in serine residues. When these residues are phosphorylated, the CTD binds to various protein factors that promote transcript maturation and modification. RNA polymerase III transcribes 5S rRNA, transfer RNA (tRNA) genes, and some small non-coding RNAs (e.g., 7SK). Transcription ends when the polymerase encounters a sequence called the terminator.

RNA Processing

While transcription of prokaryotic protein-coding genes creates messenger RNA (mRNA) that is ready for translation into protein, transcription of eukaryotic genes leaves a primary transcript of RNA (pre-mRNA), which first has to undergo a series of modifications to become a mature mRNA.

These include 5' *capping*, which is set of enzymatic reactions that add 7-methylguanosine (m^7G) to the 5' end of pre-mRNA and thus protect the RNA from degradation by exonucleases. The m^7G cap is then bound by cap binding complex heterodimer (CBC20/CBC80), which aids in mRNA export to cytoplasm and also protect the RNA from decapping.

Another modification is 3' *cleavage and polyadenylation*. They occur if polyadenylation signal sequence (5'- AAUAAA-3') is present in pre-mRNA, which is usually between protein-coding sequence and terminator. The pre-mRNA is first cleaved and then a series of ~200 adenines (A) are added to form poly(A) tail, which protects the RNA from degradation. Poly(A) tail is bound by multiple poly(A)-binding proteins (PABP) necessary for mRNA export and translation re-initiation.

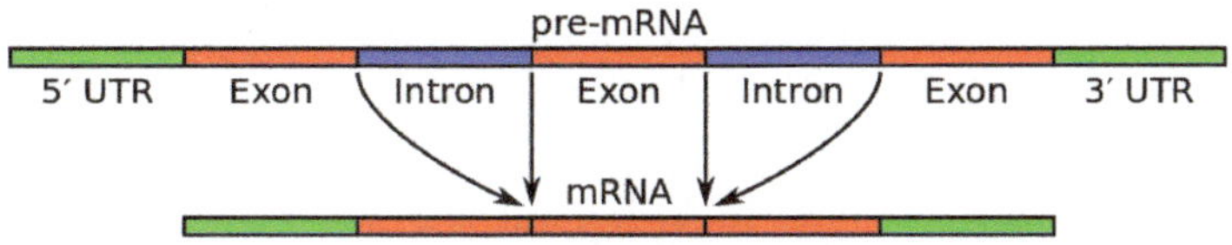

Simple illustration of exons and introns in pre-mRNA and the formation of mature mRNA by splicing. The UTRs are non-coding parts of exons at the ends of the mRNA.

A very important modification of eukaryotic pre-mRNA is *RNA splicing*. The majority of eukaryotic pre-mRNAs consist of alternating segments called exons and introns. During the process of splicing, an RNA-protein catalytical complex known as spliceosome catalyzes two transesterification reactions, which remove an intron and release it in form of lariat structure, and then splice neighbouring exons together. In certain cases, some introns or exons can be either removed or retained in mature mRNA. This so-called alternative splicing creates series of different transcripts originating from a single gene. Because these transcripts can be potentially translated into different proteins, splicing extends the complexity of eukaryotic gene expression.

Extensive RNA processing may be an evolutionary advantage made possible by the nucleus of eukaryotes. In prokaryotes, transcription and translation happen together, whilst in eukaryotes, the nuclear membrane separates the two processes, giving time for RNA processing to occur.

Non-coding RNA Maturation

In most organisms non-coding genes (ncRNA) are transcribed as precursors that undergo further processing. In the case of ribosomal RNAs (rRNA), they are often transcribed as a pre-rRNA that contains one or more rRNAs. The pre-rRNA is cleaved and modified (2′-O-methylation and pseudouridine formation) at specific sites by approximately 150 different small nucleolus-restricted RNA species, called snoRNAs. SnoRNAs associate with proteins, forming snoRNPs. While snoRNA part basepair with the target RNA and thus position the modification at a precise site, the protein part performs the catalytical reaction. In eukaryotes, in particular a snoRNP called RNase, MRP cleaves the 45S pre-rRNA into the 28S, 5.8S, and 18S rRNAs. The rRNA and RNA processing factors form large aggregates called the nucleolus.

In the case of transfer RNA (tRNA), for example, the 5' sequence is removed by RNase P, whereas the 3' end is removed by the tRNase Z enzyme and the non-templated 3' CCA tail is added by a nucleotidyl transferase. In the case of micro RNA (miRNA), miRNAs are first transcribed as primary transcripts or pri-miRNA with a cap and poly-A tail and processed to short, 70-nucleotide stem-loop structures known as pre-miRNA in the cell nucleus by the enzymes Drosha and Pasha. After being exported, it is then processed to mature miRNAs in the cytoplasm by interaction with the endonuclease Dicer, which also initiates the formation of the RNA-induced silencing complex (RISC), composed of the Argonaute protein.

Even snRNAs and snoRNAs themselves undergo series of modification before they become part of functional RNP complex. This is done either in the nucleoplasm or in the specialized compartments called Cajal bodies. Their bases are methylated or pseudouridinilated by a group of small Cajal body-specific RNAs (scaRNAs), which are structurally similar to snoRNAs.

RNA Export

In eukaryotes most mature RNA must be exported to the cytoplasm from the nucleus. While some RNAs function in the nucleus, many RNAs are transported through the nuclear pores and into the cytosol. Notably this includes all RNA types involved in protein synthesis. In some cases RNAs are additionally transported to a specific part of the cytoplasm, such as a synapse; they are then towed by motor proteins that bind through linker proteins to specific sequences (called "zipcodes") on the RNA.

Translation

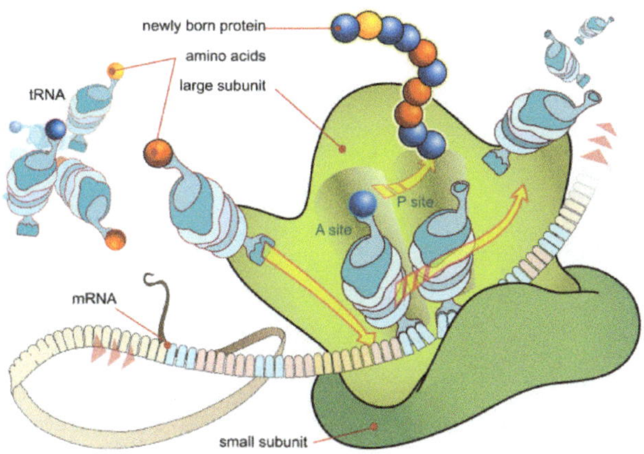

During the translation, tRNA charged with amino acid enters the ribosome and aligns with the correct mRNA triplet. Ribosome then adds amino acid to growing protein chain.

For some RNA (non-coding RNA) the mature RNA is the final gene product. In the case of messenger RNA (mRNA) the RNA is an information carrier coding for the synthesis of one or more proteins. mRNA carrying a single protein sequence (common in eukaryotes) is monocistronic whilst mRNA carrying multiple protein sequences (common in prokaryotes) is known as polycistronic.

Every mRNA consists of three parts: a 5' untranslated region (5'UTR), a protein-coding region or open reading frame (ORF), and a 3' untranslated region (3'UTR). The coding region carries information for protein synthesis encoded by the genetic code to form triplets. Each triplet of nucleotides of the coding region is called a codon and corresponds to a binding site complementary to an anticodon triplet in transfer RNA. Transfer RNAs with the same anticodon sequence always carry an identical type of amino acid. Amino acids are then chained together by the ribosome according to the order of triplets in the coding region. The ribosome helps transfer RNA to bind to messenger RNA and takes the amino acid from each transfer RNA and makes a structure-less protein out of it. Each mRNA molecule is translated into many protein molecules, on average ~2800 in mammals.

In prokaryotes translation generally occurs at the point of transcription (co-transcriptionally), often using a messenger RNA that is still in the process of being created. In eukaryotes translation can occur in a variety of regions of the cell depending on where the protein being written is supposed to be. Major locations are the cytoplasm for soluble cytoplasmic proteins and the membrane of the endoplasmic reticulum for proteins that are for export from the cell or insertion into a cell membrane. Proteins that are supposed to be expressed at the endoplasmic reticulum are recognised part-way through the translation process. This is governed by the signal recognition particle—a protein that binds to the ribosome and directs it to the endoplasmic reticulum when it finds a signal peptide on the growing (nascent) amino acid chain. Translation is the communication of the meaning of a source-language text by means of an equivalent target-language text

Folding

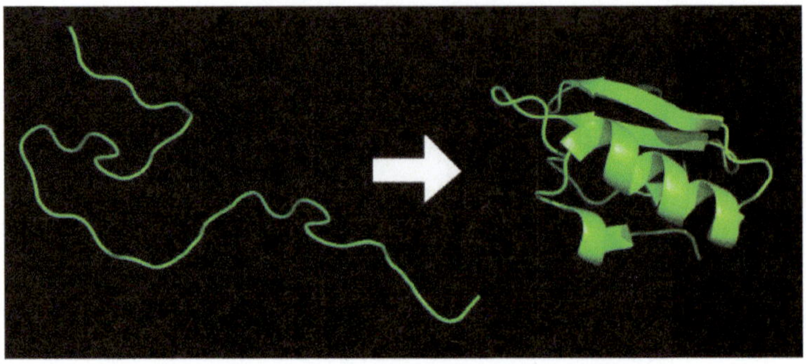

Protein before (left) and after (right) folding.

The polypeptide folds into its characteristic and functional three-dimensional structure from a random coil. Each protein exists as an unfolded polypeptide or random coil when translated from a sequence of mRNA into a linear chain of amino acids. This polypeptide lacks any developed three-dimensional structure (the left hand side of the neighboring figure). Amino acids interact with each other to produce a well-defined three-dimensional structure, the folded protein (the right hand side of the figure) known as the native state. The resulting three-dimensional structure is determined by the amino acid sequence (Anfinsen's dogma).

The correct three-dimensional structure is essential to function, although some parts of functional proteins may remain unfolded Failure to fold into the intended shape usually produces inactive proteins with different properties including toxic prions. Several neurodegenerative and other diseases are believed to result from the accumulation of *misfolded* proteins. Many allergies are caused by the folding of the proteins, for the immune system does not produce antibodies for certain protein structures.

Enzymes called chaperones assist the newly formed protein to attain (fold into) the 3-dimensional structure it needs to function. Similarly, RNA chaperones help RNAs attain their functional shapes. Assisting protein folding is one of the main roles of the endoplasmic reticulum in eukaryotes.

Translocation

Secretory proteins of eukaryotes or prokaryotes must be translocated to enter the secretory pathway. Newly synthesized proteins are directed to the eukaryotic Sec61 or prokaryotic SecYEG translocation channel by signal peptides. The efficiency of protein secretion in eukaryotes is very dependent on the signal peptide which has been used.

Protein Transport

Many proteins are destined for other parts of the cell than the cytosol and a wide range of signalling sequences or (signal peptides) are used to direct proteins to where they are supposed to be. In prokaryotes this is normally a simple process due to limited compartmentalisation of the cell. However, in eukaryotes there is a great variety of different targeting processes to ensure the protein arrives at the correct organelle.

Not all proteins remain within the cell and many are exported, for example, digestive enzymes, hormones and extracellular matrix proteins. In eukaryotes the export pathway is well developed and the main mechanism for the export of these proteins is translocation to the endoplasmic reticulum, followed by transport via the Golgi apparatus.

Regulation of Gene Expression

The patchy colours of a tortoiseshell cat are the result of different levels of expression of pigmentation genes in different areas of the skin.

Regulation of gene expression refers to the control of the amount and timing of appearance of the functional product of a gene. Control of expression is vital to allow a cell to produce the gene products it needs when it needs them; in turn, this gives cells the flexibility to adapt to a variable environment, external signals, damage to the cell, and other stimuli. More generally, gene regulation gives the cell control over all structure and function, and is the basis for cellular differentiation, morphogenesis and the versatility and adaptability of any organism.

Numerous terms are used to describe types of genes depending on how they are regulated; these include:

- A constitutive gene is a gene that is transcribed continually as opposed to a facultative gene, which is only transcribed when needed.
- A *housekeeping gene* is a gene that is required to maintain basic cellular function and so is typically expressed in all cell types of an organism. Examples include actin, GAPDH and ubiquitin. Some housekeeping genes are transcribed at a relatively constant rate and these genes can be used as a reference point in experiments to measure the expression rates of other genes.
- A facultative gene is a gene only transcribed when needed as opposed to a constitutive gene.
- An inducible gene is a gene whose expression is either responsive to environmental change or dependent on the position in the cell cycle.

Any step of gene expression may be modulated, from the DNA-RNA transcription step to post-translational modification of a protein. The stability of the final gene product, whether it is RNA or protein, also contributes to the expression level of the gene—an unstable product results in a low expression level. In general gene expression is regulated through changes in the number and type of interactions between molecules that collectively influence transcription of DNA and translation of RNA.

Some simple examples of where gene expression is important are:

- Control of insulin expression so it gives a signal for blood glucose regulation.
- X chromosome inactivation in female mammals to prevent an "overdose" of the genes it contains.
- Cyclin expression levels control progression through the eukaryotic cell cycle.

Transcriptional Regulation

Regulation of transcription can be broken down into three main routes of influence; genetic (direct interaction of a control factor with the gene), modulation interaction of a control factor with the transcription machinery and epigenetic (non-sequence changes in DNA structure that influence transcription).

Direct interaction with DNA is the simplest and the most direct method by which a protein changes transcription levels. Genes often have several protein binding sites around the coding region with the specific function of regulating transcription. There are many classes of regulatory DNA binding sites known as enhancers, insulators and silencers. The mechanisms for regulating transcription are very varied, from blocking key binding sites on the DNA for RNA polymerase to acting as an activator and promoting transcription by assisting RNA polymerase binding.

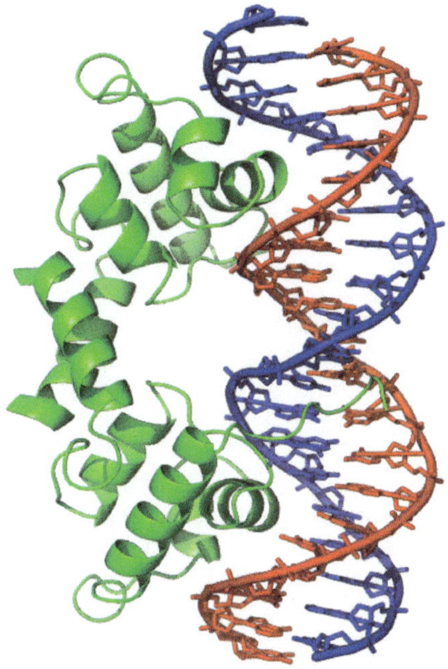

The lambda repressor transcription factor (green) binds as a dimer to major groove of DNA target (red and blue) and disables initiation of transcription. From PDB: 1LMB.

The activity of transcription factors is further modulated by intracellular signals causing protein post-translational modification including phosphorylated, acetylated, or glycosylated. These changes influence a transcription factor's ability to bind, directly or indirectly, to promoter DNA, to recruit RNA polymerase, or to favor elongation of a newly synthesized RNA molecule.

The nuclear membrane in eukaryotes allows further regulation of transcription factors by the duration of their presence in the nucleus, which is regulated by reversible changes in their structure and by binding of other proteins. Environmental stimuli or endocrine signals may cause modification of regulatory proteins eliciting cascades of intracellular signals, which result in regulation of gene expression.

More recently it has become apparent that there is a significant influence of non-DNA-sequence specific effects on transcription. These effects are referred to as epigenetic and involve the higher order structure of DNA, non-sequence specific DNA binding proteins and chemical modification of DNA. In general epigenetic effects alter the accessibility of DNA to proteins and so modulate transcription.

DNA methylation is a widespread mechanism for epigenetic influence on gene expression and is seen in bacteria and eukaryotes and has roles in heritable transcription silencing and transcription regulation. In eukaryotes the structure of chromatin, controlled by the histone code, regulates access to DNA with significant impacts on the expression of genes in euchromatin and heterochromatin areas.

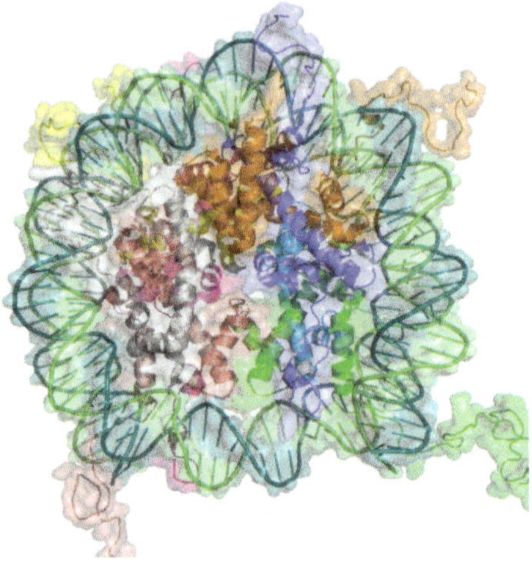

In eukaryotes, DNA is organized in form of nucleosomes. Note how the DNA (blue and green) is tightly wrapped around the protein core made of histone octamer (ribbon coils), restricting access to the DNA. From PDB: 1KX5.

Transcriptional Regulation in Cancer

The majority of gene promoters contain a CpG island with numerous CpG sites. When many of a gene's promoter CpG sites are methylated the gene becomes silenced. Colorectal cancers typically have 3 to 6 driver mutations and 33 to 66 hitchhiker or passenger mutations. However, transcriptional silencing may be of more importance than mutation in causing progression to cancer. For example, in colorectal cancers about 600 to 800 genes are transcriptionally silenced by CpG island methylation. Transcriptional repression in cancer can also occur by other epigenetic mechanisms, such as altered expression of microRNAs. In breast cancer, transcriptional repression of BRCA1 may occur more frequently by over-expressed microRNA-182 than by hypermethylation of the BRCA1 promoter.

Post-transcriptional Regulation

In eukaryotes, where export of RNA is required before translation is possible, nuclear export is thought to provide additional control over gene expression. All transport in and out of the nucleus is via the nuclear pore and transport is controlled by a wide range of importin and exportin proteins.

Expression of a gene coding for a protein is only possible if the messenger RNA carrying the code survives long enough to be translated. In a typical cell, an RNA molecule is only stable if specifically protected from degradation. RNA degradation has particular importance in regulation of expression in eukaryotic cells where mRNA has to travel significant distances before being translated. In eukaryotes, RNA is stabilised by certain post-transcriptional modifications, particularly the 5' cap and poly-adenylated tail.

Intentional degradation of mRNA is used not just as a defence mechanism from foreign RNA (normally from viruses) but also as a route of mRNA *destabilisation*. If an mRNA molecule has a complementary sequence to a small interfering RNA then it is targeted for destruction via the RNA interference pathway.

Three Prime Untranslated Regions and MicroRNAs

Three prime untranslated regions (3'UTRs) of messenger RNAs (mRNAs) often contain regulatory sequences that post-transcriptionally influence gene expression. Such 3'-UTRs often contain both binding sites for microRNAs (miRNAs) as well as for regulatory proteins. By binding to specific sites within the 3'-UTR, miRNAs can decrease gene expression of various mRNAs by either inhibiting translation or directly causing degradation of the transcript. The 3'-UTR also may have silencer regions that bind repressor proteins that inhibit the expression of a mRNA.

The 3'-UTR often contains microRNA response elements (MREs). MREs are sequences to which miRNAs bind. These are prevalent motifs within 3'-UTRs. Among all regulatory motifs within the 3'-UTRs (e.g. including silencer regions), MREs make up about half of the motifs.

As of 2014, the miRBase web site, an archive of miRNA sequences and annotations, listed 28,645 entries in 233 biologic species. Of these, 1,881 miRNAs were in annotated human miRNA loci. miRNAs were predicted to have an average of about four hundred target mRNAs (affecting expression of several hundred genes). Freidman et al. estimate that >45,000 miRNA target sites within human mRNA 3'UTRs are conserved above background levels, and >60% of human protein-coding genes have been under selective pressure to maintain pairing to miRNAs.

Direct experiments show that a single miRNA can reduce the stability of hundreds of unique mRNAs. Other experiments show that a single miRNA may repress the production of hundreds of proteins, but that this repression often is relatively mild (less than 2-fold).

The effects of miRNA dysregulation of gene expression seem to be important in cancer. For instance, in gastrointestinal cancers, nine miRNAs have been identified as epigenetically altered and effective in down regulating DNA repair enzymes.

The effects of miRNA dysregulation of gene expression also seem to be important in neuropsychiatric disorders, such as schizophrenia, bipolar disorder, major depression, Parkinson's disease, Alzheimer's disease and autism spectrum disorders.

Translational Regulation

Direct regulation of translation is less prevalent than control of transcription or mRNA stability but is occasionally used. Inhibition of protein translation is a major target for toxins and antibiotics, so they can kill a cell by overriding its normal gene expression control. Protein synthesis inhibitors include the antibiotic neomycin and the toxin ricin.

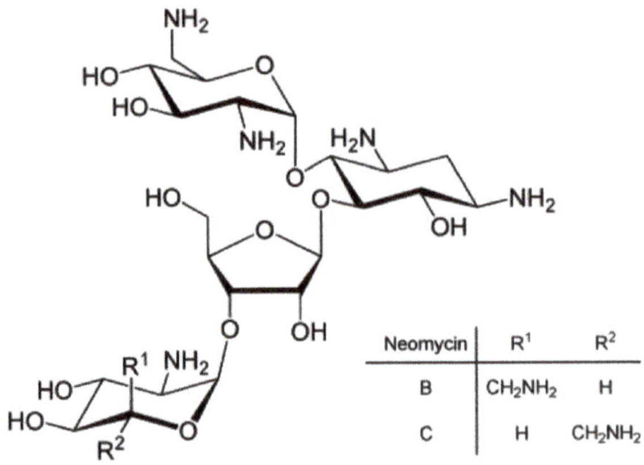

Neomycin	R^1	R^2
B	CH_2NH_2	H
C	H	CH_2NH_2

Neomycin is an example of a small molecule that reduces expression of all protein genes inevitably leading to cell death; it thus acts as an antibiotic.

Protein Degradation

Once protein synthesis is complete, the level of expression of that protein can be reduced by protein degradation. There are major protein degradation pathways in all prokaryotes and eukaryotes, of which the proteasome is a common component. An unneeded or damaged protein is often labeled for degradation by addition of ubiquitin.

Measurement

Measuring gene expression is an important part of many life sciences, as the ability to quantify the level at which a particular gene is expressed within a cell, tissue or organism can provide a lot of valuable information. For example, measuring gene expression can:

- Identify viral infection of a cell (viral protein expression).
- Determine an individual's susceptibility to cancer (oncogene expression).
- Find if a bacterium is resistant to penicillin (beta-lactamase expression).

Similarly, the analysis of the location of protein expression is a powerful tool, and this can be done on an organismal or cellular scale. Investigation of localization is particularly important for the study of development in multicellular organisms and as an indicator of protein function in single cells. Ideally, measurement of expression is done by detecting the final gene product (for many genes, this is the protein); however, it is often easier to detect one of the precursors, typically mRNA and to infer gene-expression levels from these measurements.

mRNA Quantification

Levels of mRNA can be quantitatively measured by northern blotting, which provides size and sequence information about the mRNA molecules. A sample of RNA is separated on an agarose gel and hybridized to a radioactively labeled RNA probe that is complementary to the target sequence. The radiolabeled RNA is then detected by an autoradiograph. Because the use of radioactive reagents makes the procedure time consuming and potentially dangerous, alternative labeling and detection methods, such as digoxigenin and biotin chemistries, have been developed. Perceived disadvantages of Northern blotting are that large quantities of RNA are required and that quantification may not be completely accurate, as it involves measuring band strength in an image of a gel. On the other hand, the additional mRNA size information from the Northern blot allows the discrimination of alternately spliced transcripts.

Another approach for measuring mRNA abundance is RT-qPCR. In this technique, reverse transcription is followed by quantitative PCR. Reverse transcription first generates a DNA template from the mRNA; this single-stranded template is called cDNA. The cDNA template is then amplified in the quantitative step, during which the fluorescence emitted by labeled hybridization probes or intercalating dyes changes as the DNA amplification process progresses. With a carefully constructed standard curve, qPCR can produce an absolute measurement of the number of copies of original mRNA, typically in units of copies per nanolitre of homogenized tissue or copies per cell. qPCR is very sensitive (detection of a single mRNA molecule is theoretically possible), but can be expensive depending on the type of reporter used; fluorescently labeled oligonucleotide probes are more expensive than non-specific intercalating fluorescent dyes.

For expression profiling, or high-throughput analysis of many genes within a sample, quantitative PCR may be performed for hundreds of genes simultaneously in the case of low-density arrays. A second approach is the hybridization microarray. A single array or "chip" may contain probes to determine transcript levels for every known gene in the genome of one or more organisms. Alternatively, "tag based" technologies like Serial analysis of gene expression (SAGE) and RNA-Seq, which can provide a relative measure of the cellular concentration of different mRNAs, can be used. An advantage of tag-based methods is the "open architecture", allowing for the exact measurement of any transcript, with a known or unknown sequence. Next-generation sequencing (NGS) such as RNA-Seq is another approach, producing vast quantities of sequence data that can be matched to a reference genome. Although NGS is comparatively time-consuming, expensive, and resource-intensive, it can identify single-nucleotide polymorphisms, splice-variants, and novel genes, and can also be used to profile expression in organisms for which little or no sequence information is available.

Protein Quantification

For genes encoding proteins, the expression level can be directly assessed by a number of methods with some clear analogies to the techniques for mRNA quantification.

The most commonly used method is to perform a Western blot against the protein of interest—this gives information on the size of the protein in addition to its identity. A sample (often cellular lysate) is separated on a polyacrylamide gel, transferred to a membrane and then probed with an antibody to the protein of interest. The antibody can either be conjugated to a fluorophore or to horseradish peroxidase for imaging and/ or quantification. The gel-based nature of this assay makes quantification less accurate, but it has the advantage of being able to identify later modifications to the protein, for example proteolysis or ubiquitination, from changes in size.

Localisation

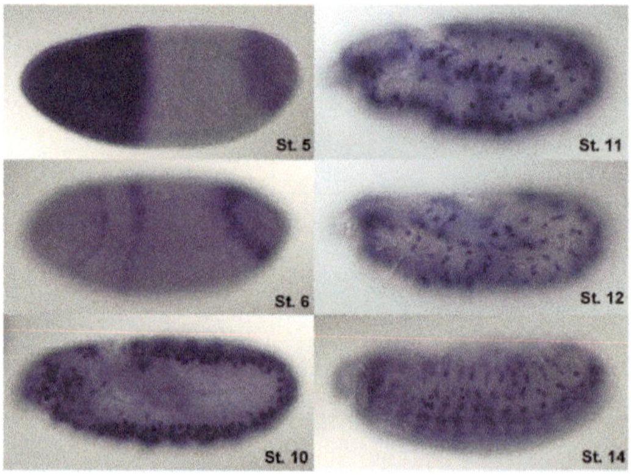

In situ-hybridization of Drosophila embryos at different developmental stages for the mRNA responsible for the expression of hunchback. High intensity of blue color marks places with high hunchback mRNA quantity.

Analysis of expression is not limited to quantification; localisation can also be determined. mRNA can be detected with a suitably labelled complementary mRNA strand and protein can be detected via labelled antibodies. The probed sample is then observed by microscopy to identify where the mRNA or protein is.

By replacing the gene with a new version fused to a green fluorescent protein (or similar) marker, expression may be directly quantified in live cells. This is done by imaging using a fluorescence microscope. It is very difficult to clone a GFP-fused protein into its native location in the genome without affecting expression levels so this method often cannot be used to measure endogenous gene expression. It is, however, widely used to measure the expression of a gene artificially introduced into the cell, for example via an expression vector. It is important to note that by fusing a target protein to a fluorescent reporter the protein's behavior, including its cellular localization and expression level, can be significantly changed.

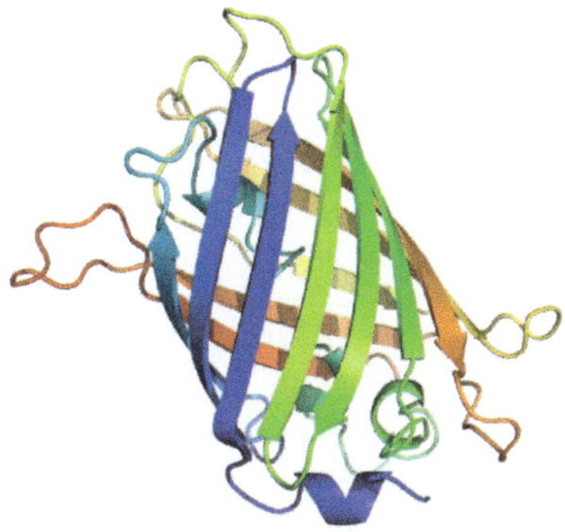

The three-dimensional structure of green fluorescent protein. The residues in the centre of the "barrel" are responsible for production of green light after exposing to higher energetic blue light. From PDB: 1EMA.

The enzyme-linked immunosorbent assay works by using antibodies immobilised on a microtiter plate to capture proteins of interest from samples added to the well. Using a detection antibody conjugated to an enzyme or fluorophore the quantity of bound protein can be accurately measured by fluorometric or colourimetric detection. The detection process is very similar to that of a Western blot, but by avoiding the gel steps more accurate quantification can be achieved.

Expression System

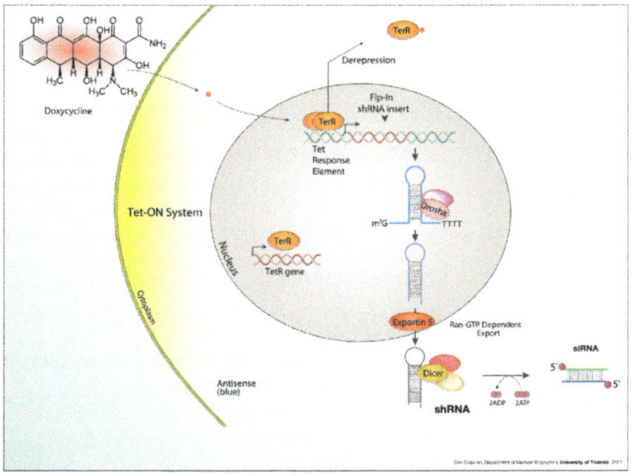

Tet-ON inducible shRNA system

An expression system is a system specifically designed for the production of a gene product of choice. This is normally a protein although may also be RNA, such as tRNA or a ribozyme. An expression system consists of a gene, normally encoded by DNA, and the molecular machinery required to transcribe the DNA into mRNA and translate the mRNA into protein using the reagents provided. In the broadest sense this includes every living cell but the term is more normally used to refer to expression as a laboratory tool. An expression system is therefore often artificial in some manner. Expression systems are, however, a fundamentally natural process. Viruses are an excellent example where they replicate by using the host cell as an expression system for the viral proteins and genome.

Inducible Expression

Doxycycline is also used in "Tet-on" and "Tet-off" tetracycline controlled transcriptional activation to regulate transgene expression in organisms and cell cultures.

In Nature

In addition to these biological tools, certain naturally observed configurations of DNA (genes, promoters, enhancers, repressors) and the associated machinery itself are referred to as an expression system. This term is normally used in the case where a gene or set of genes is switched on under well defined conditions, for example, the simple repressor switch expression system in Lambda phage and the lac operator system in bacteria. Several natural expression systems are directly used or modified and used for artificial expression systems such as the Tet-on and Tet-off expression system.

Gene Networks

Genes have sometimes been regarded as nodes in a network, with inputs being proteins such as transcription factors, and outputs being the level of gene expression. The node itself performs a function, and the operation of these functions have been interpreted as performing a kind of information processing within cells and determines cellular behavior.

Gene networks can also be constructed without formulating an explicit causal model. This is often the case when assembling networks from large expression data sets. Covariation and correlation of expression is computed across a large sample of cases and measurements (often transcriptome or proteome data). The source of variation can be either experimental or natural (observational). There are several ways to construct gene expression networks, but one common approach is to compute a matrix of all pair-wise correlations of expression across conditions, time points, or individuals and convert the matrix (after thresholding at some cut-off value) into a graphical representation in which nodes represent genes, transcripts, or proteins and edges connecting these nodes represent the strength of association. Weighted correlation network analysis involves weighted networks defined by soft-thresholding the pairwise correlations among variables (e.g. measures of transcript abundance).

Regulation of Gene Expression

Regulation of gene expression includes a wide range of mechanisms that are used by cells to increase or decrease the production of specific gene products (protein or RNA), and is informally termed *gene regulation*. Sophisticated programs of gene expression are widely observed in biology, for example to trigger developmental pathways, respond to environmental stimuli, or adapt to new food sources. Virtually any step of gene expression can be modulated, from transcriptional initiation, to RNA processing, and to the post-translational modification of a protein. Often, one gene regulator controls another, and so on, in a gene regulatory network.

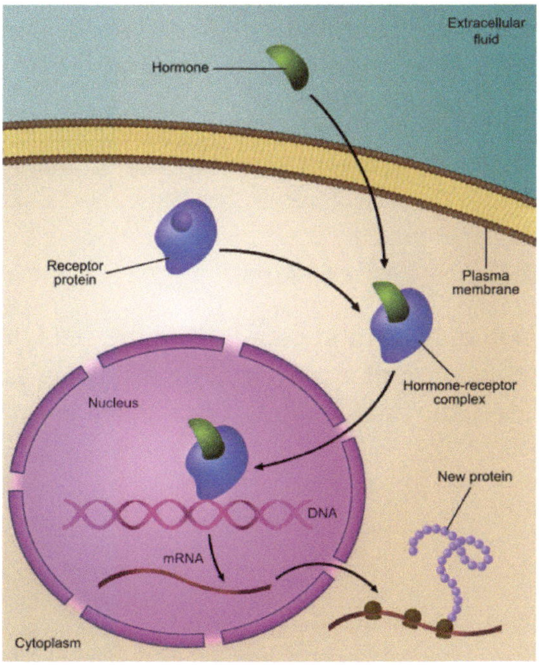

Regulation of gene expression by a hormone receptor

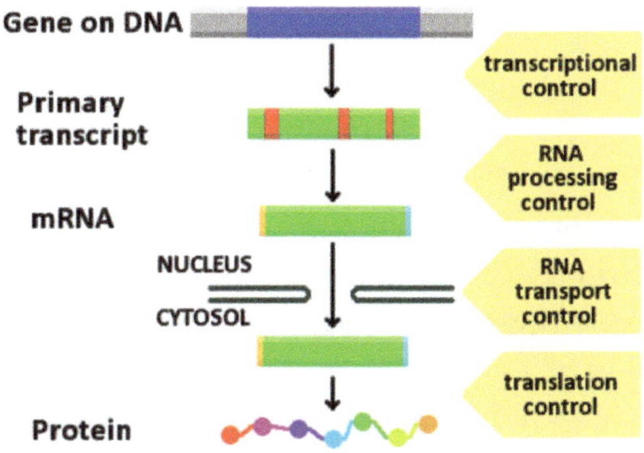

Diagram showing at which stages in the DNA-mRNA-protein pathway expression can be controlled

Gene regulation is essential for viruses, prokaryotes and eukaryotes as it increases the versatility and adaptability of an organism by allowing the cell to express protein when needed. Although as early as 1951, Barbara McClintock showed interaction between two genetic loci, Activator (*Ac*) and Dissociator (*Ds*), in the color formation of maize seeds, the first discovery of a gene regulation system is widely considered to be the identification in 1961 of the *lac* operon, discovered by Jacques Monod, in which some enzymes involved in lactose metabolism are expressed by *E. coli* only in the presence of lactose and absence of glucose.

In multicellular organisms, gene regulation drives cellular differentiation and morphogenesis in the embryo, leading to the creation of different cell types that possess different gene expression profiles from the same genome sequence. This explains how evolution actually works at a molecular level, and is central to the science of evolutionary developmental biology ("evo-devo").

The initiating event leading to a change in gene expression includes activation or deactivation of receptors.

Regulated Stages of Gene Expression

Any step of gene expression may be modulated, from the DNA-RNA transcription step to post-translational modification of a protein. The following is a list of stages where gene expression is regulated, the most extensively utilised point is Transcription Initiation:

- Chromatin domains
- Transcription
- Post-transcriptional modification
- RNA transport
- Translation
- mRNA degradation

Modification of DNA

In eukaryotes, the accessibility of large regions of DNA can depend on its chromatin structure, which can be altered as a result of histone modifications directed by DNA methylation, ncRNA, or DNA-binding protein. Hence these modifications may up or down regulate the expression of a gene. Some of these modifications that regulate gene expression are inheritable and are referred to as epigenetic regulation.

Structural

Transcription of DNA is dictated by its structure. In general, the density of its packing is indicative of the frequency of transcription. Octameric protein complexes called nucleosomes are responsible for the amount of supercoiling of DNA, and these complexes can be temporarily modified by processes such as phosphorylation or more permanently modified by processes such as methylation. Such modifications are considered to be responsible for more or less permanent changes in gene expression levels.

Chemical

Methylation of DNA is a common method of gene silencing. DNA is typically methylated by methyltransferase enzymes on cytosine nucleotides in a CpG dinucleotide sequence (also called "CpG islands" when densely clustered). Analysis of the pattern of methylation in a given region of DNA (which can be a promoter) can be achieved through a method called bisulfite mapping. Methylated cytosine residues are unchanged by the treatment, whereas unmethylated ones are changed to uracil. The differences are analyzed by DNA sequencing or by methods developed to quantify SNPs, such as Pyrosequencing (Biotage) or MassArray (Sequenom), measuring the relative amounts of C/T at the CG dinucleotide. Abnormal methylation patterns are thought to be involved in oncogenesis.

Histone acetylation is also an important process in transcription. Histone acetyltransferase enzymes (HATs) such as CREB-binding protein also dissociate the DNA from the histone complex, allowing transcription to proceed. Often, DNA methylation and histone deacetylation work together in gene silencing. The combination of the two seems to be a signal for DNA to be packed more densely, lowering gene expression.

Regulation of Transcription

Regulation of transcription thus controls when transcription occurs and how much RNA is created. Transcription of a gene by RNA polymerase can be regulated by several mechanisms. Specificity factors alter the specificity of RNA polymerase for a given promoter or set of promoters, making it more or less likely to bind to them (i.e., sigma factors used in prokaryotic transcription). Repressors bind to the Operator, coding sequences on the DNA strand that are close to or overlapping the promoter region, impeding RNA polymerase's progress along the strand, thus impeding the expression of the gene.The image to the right demonstrates regulation by a repressor in the lac operon. General transcription factors position RNA polymerase at the start of a protein-coding sequence and then release the polymerase to transcribe the mRNA. Activators enhance the interaction between RNA polymerase and a particular promoter, encouraging the expression of the gene. Activators do this by increasing the attraction of RNA polymerase for the promoter, through interactions with subunits of the RNA polymerase or indirectly by changing the structure of the DNA. Enhancers are sites on the DNA helix that are bound by activators in order to loop the DNA bringing a specific promoter to the initiation complex. Enhancers are much more common in eukaryotes than prokaryotes, where only a few examples exist (to date). Silencers are regions of DNA sequences that, when bound by particular transcription factors, can silence expression of the gene.

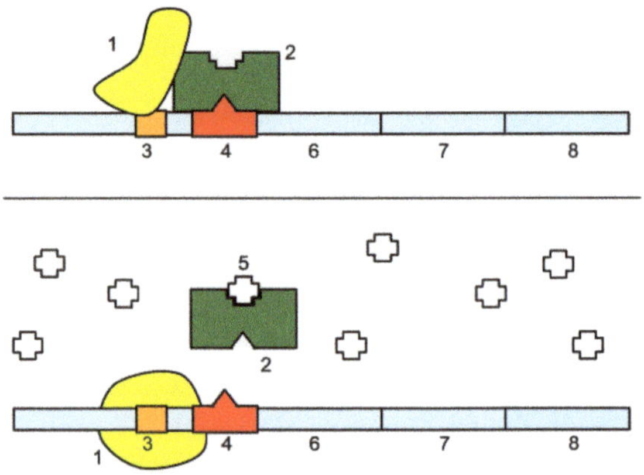

1: RNA Polymerase, *2*: Repressor, *3*: Promoter, *4*: Operator, *5*: Lactose, *6*: lacZ, *7*: lacY, *8*: lacA. Top: The gene is essentially turned off. There is no lactose to inhibit the repressor, so the repressor binds to the operator, which obstructs the RNA polymerase from binding to the promoter and making lactase. Bottom: The gene is turned on. Lactose is inhibiting the repressor, allowing the RNA polymerase to bind with the promoter, and express the genes, which synthesize lactase. Eventually, the lactase will digest all of the lactose, until there is none to bind to the repressor. The repressor will then bind to the operator, stopping the manufacture of lactase.

Regulation of Transcription in Cancer

In vertebrates, the majority of gene promoters contain a CpG island with numerous CpG sites. When many of a gene's promoter CpG sites are methylated the gene becomes silenced. Colorectal cancers typically have 3 to 6 driver mutations and 33 to 66 hitchhiker or passenger mutations. However, transcriptional silencing may be of more importance than mutation in causing progression to cancer. For example, in colorectal cancers about 600 to 800 genes are transcriptionally silenced by CpG island methylation. Transcriptional repression in cancer can also occur by other epigenetic mechanisms, such as altered expression of microRNAs. In breast cancer, transcriptional repression of BRCA1 may occur more frequently by over-expressed microRNA-182 than by hypermethylation of the BRCA1 promoter.

Post-transcriptional Regulation

After the DNA is transcribed and mRNA is formed, there must be some sort of regulation on how much the mRNA is translated into proteins. Cells do this by modulating the capping, splicing, addition of a Poly(A) Tail, the sequence-specific nuclear export rates, and, in several contexts, sequestration of the RNA transcript. These processes occur in eukaryotes but not in prokaryotes. This modulation is a result of a protein or transcript that, in turn, is regulated and may have an affinity for certain sequences.

Three Prime Untranslated Regions and MicroRNAs

Three prime untranslated regions (3'-UTRs) of messenger RNAs (mRNAs) often contain regulatory sequences that post-transcriptionally influence gene expression. Such

3'-UTRs often contain both binding sites for microRNAs (miRNAs) as well as for regulatory proteins. By binding to specific sites within the 3'-UTR, miRNAs can decrease gene expression of various mRNAs by either inhibiting translation or directly causing degradation of the transcript. The 3'-UTR also may have silencer regions that bind repressor proteins that inhibit the expression of a mRNA.

The 3'-UTR often contains miRNA response elements (MREs). MREs are sequences to which miRNAs bind. These are prevalent motifs within 3'-UTRs. Among all regulatory motifs within the 3'-UTRs (e.g. including silencer regions), MREs make up about half of the motifs.

As of 2014, the miRBase web site, an archive of miRNA sequences and annotations, listed 28,645 entries in 233 biologic species. Of these, 1,881 miRNAs were in annotated human miRNA loci. miRNAs were predicted to have an average of about four hundred target mRNAs (affecting expression of several hundred genes). Freidman et al. estimate that >45,000 miRNA target sites within human mRNA 3'-UTRs are conserved above background levels, and >60% of human protein-coding genes have been under selective pressure to maintain pairing to miRNAs.

Direct experiments show that a single miRNA can reduce the stability of hundreds of unique mRNAs. Other experiments show that a single miRNA may repress the production of hundreds of proteins, but that this repression often is relatively mild (less than 2-fold).

The effects of miRNA dysregulation of gene expression seem to be important in cancer. For instance, in gastrointestinal cancers, a 2015 paper identified nine miRNAs as epigenetically altered and effective in down-regulating DNA repair enzymes.

The effects of miRNA dysregulation of gene expression also seem to be important in neuropsychiatric disorders, such as schizophrenia, bipolar disorder, major depressive disorder, Parkinson's disease, Alzheimer's disease and autism spectrum disorders.

Regulation of Translation

The translation of mRNA can also be controlled by a number of mechanisms, mostly at the level of initiation. Recruitment of the small ribosomal subunit can indeed be modulated by mRNA secondary structure, antisense RNA binding, or protein binding. In both prokaryotes and eukaryotes, a large number of RNA binding proteins exist, which often are directed to their target sequence by the secondary structure of the transcript, which may change depending on certain conditions, such as temperature or presence of a ligand (aptamer). Some transcripts act as ribozymes and self-regulate their expression.

Examples of Gene Regulation

- Enzyme induction is a process in which a molecule (e.g., a drug) induces (i.e., initiates or enhances) the expression of an enzyme.

- The induction of heat shock proteins in the fruit fly *Drosophila melanogaster*.
- The Lac operon is an interesting example of how gene expression can be regulated.
- Viruses, despite having only a few genes, possess mechanisms to regulate their gene expression, typically into an early and late phase, using collinear systems regulated by anti-terminators (lambda phage) or splicing modulators (HIV).
- GAL4 is a transcriptional activator that controls the expression of GAL1, GAL7, and GAL10 (all of which code for the metabolic of galactose in yeast). The GAL4/UAS system has been used in a variety of organisms across various phyla to study gene expression.

Developmental Biology

A large number of studied regulatory systems come from developmental biology. Examples include:

- The colinearity of the Hox gene cluster with their nested antero-posterior patterning.
- It has been speculated that pattern generation of the hand (digits - interdigits) The gradient of Sonic hedgehog (secreted inducing factor) from the zone of polarizing activity in the limb, which creates a gradient of active Gli3, which activates Gremlin, which inhibits BMPs also secreted in the limb, resulting in the formation of an alternating pattern of activity as a result of this reaction-diffusion system.
- Somitogenesis is the creation of segments (somites) from a uniform tissue (Pre-somitic Mesoderm). They are formed sequentially from anterior to posterior. This is achieved in amniotes possibly by means of two opposing gradients, Retinoic acid in the anterior (wavefront) and Wnt and Fgf in the posterior, coupled to an oscillating pattern (segmentation clock) composed of FGF + Notch and Wnt in antiphase.
- Sex determination in the soma of a Drosophila requires the sensing of the ratio of autosomal genes to sex chromosome-encoded genes, which results in the production of sexless splicing factor in females, resulting in the female isoform of doublesex.

Circuitry

Up-regulation and Down-regulation

Up-regulation is a process that occurs within a cell triggered by a signal (originating internal or external to the cell), which results in increased expression of one or more genes and as a result the protein(s) encoded by those genes. Conversely, down-regulation is a process resulting in decreased gene and corresponding protein expression.

- Up-regulation occurs, for example, when a cell is deficient in some kind of receptor. In this case, more receptor protein is synthesized and transported to the membrane of the cell and, thus, the sensitivity of the cell is brought back to normal, reestablishing homeostasis.
- Down-regulation occurs, for example, when a cell is overstimulated by a neurotransmitter, hormone, or drug for a prolonged period of time, and the expression of the receptor protein is decreased in order to protect the cell.

Inducible Vs. Repressible Systems

Gene Regulation can be summarized by the response of the respective system:

- Inducible systems - An inducible system is off unless there is the presence of some molecule (called an inducer) that allows for gene expression. The molecule is said to "induce expression". The manner by which this happens is dependent on the control mechanisms as well as differences between prokaryotic and eukaryotic cells.
- Repressible systems - A repressible system is on except in the presence of some molecule (called a corepressor) that suppresses gene expression. The molecule is said to "repress expression". The manner by which this happens is dependent on the control mechanisms as well as differences between prokaryotic and eukaryotic cells.

The GAL4/UAS system is an example of both an inducible and repressible system. GAL4 binds an upstream activation sequence (UAS) to activate the transcription of the GAL1/GAL7/GAL10 cassette. On the other hand, a MIG1 response to the presence of glucose can inhibit GAL4 and therefore stop the expression of the GAL1/GAL7/GAL10 cassette.

Theoretical Circuits

- Repressor/Inducer: an activation of a sensor results in the change of expression of a gene.
- Negative feedback: the gene product downregulates its own production directly or indirectly, which can result in
    - keeping transcript levels constant/proportional to a factor

    - inhibition of run-away reactions when coupled with a positive feedback loop
    - creating an oscillator by taking advantage in the time delay of transcription and translation, given that the mRNA and protein half-life is shorter
- Positive feedback: the gene product upregulates its own production directly or indirectly, which can result in
    - signal amplification
    - bistable switches when two genes inhibit each other and both have positive feedback
    - pattern generation

Study Methods

In general, most experiments investigating differential expression used whole cell extracts of RNA, called steady-state levels, to determine which genes changed and by how much they did. These are, however, not informative of where the regulation has occurred and may actually mask conflicting regulatory processes, but it is still the most commonly analysed (quantitative PCR and DNA microarray).

When studying gene expression, there are several methods to look at the various stages. In eukaryotes these include:

- The local chromatin environment of the region can be determined by ChIP-chip analysis by pulling down RNA Polymerase II, Histone 3 modifications, Trithorax-group protein, Polycomb-group protein, or any other DNA-binding element to which a good antibody is available.
- Epistatic interactions can be investigated by synthetic genetic array analysis.
- Due to post-transcriptional regulation, transcription rates and total RNA levels differ significantly. To measure the transcription rates nuclear run-on assays can be done and newer high-throughput methods are being developed, using thiol labelling instead of radioactivity.
- Only 5% of the RNA polymerised in the nucleus actually exits, and not only introns, abortive products, and non-sense transcripts are degradated. Therefore, the differences in nuclear and cytoplasmic levels can be seen by separating the two fractions by gentle lysis.
- Alternative splicing can be analysed with a splicing array or with a tiling array.
- All in vivo RNA is complexed as RNPs. The quantity of transcripts bound to

specific protein can be also analysed by RIP-Chip. For example, DCP2 will give an indication of sequestered protein; ribosome-bound gives and indication of transcripts active in transcription (although it should be noted that a more dated method, called polysome fractionation, is still popular in some labs).

- Protein levels can be analysed by Mass spectrometry, which can be compared only to quantitative PCR data, as microarray data is relative and not absolute.
- RNA and protein degradation rates are measured by means of transcription inhibitors (actinomycin D or α-amanitin) or translation inhibitors (Cycloheximide), respectively.

References

- Reissman, S. (2014). "Cell penetration: scope and limitations by the application of cell-penetrating peptides". Journal of Peptide Science. 20 (10): 760–784. doi:10.1002/psc.2672
- Ryan KJ, Ray CG (editors) (2004). Sherris Medical Microbiology (4th ed.). McGraw Hill. pp. 555–62. ISBN 0-8385-8529-9
- "Amgen Presents Interim Overall Survival Data From Phase 3 Study Of Talimogene Laherparepvec In Patients With Metastatic Melanoma" (Press release). November 18, 2013. Retrieved 30 October 2015
- Corey L, Wald A (2009). "Maternal and Neonatal HSV Infections". New England Journal of Medicine. 361 (14): 1376–85. doi:10.1056/NEJMra0807633. PMC 2780322 . PMID 19797284
- Jeremy M. Berg, John L. Tymoczko, Lubert Stryer; Web content by Neil D. Clarke (2002). "3. Protein Structure and Function". Biochemistry. San Francisco: W. H. Freeman. ISBN 0-7167-4684-0. CS1 maint: Multiple names: authors list (link)
- "Biological Efficacy Study of HerpV Vaccine With QS-21 to Treat Subjects With Recurrent Genital Herpes". Retrieved 31 August 2016
- khtar J, Shukla D (2009). "Viral entry mechanisms: Cellular and viral mediators of herpes simplex virus entry". FEBS Journal. 276 (24): 7228–7236. doi:10.1111/j.1742-4658.2009.07402.x. PMC 2801626 . PMID 19878306
- Corporation, NanoBio. "NanoBio's Genital Herpes Vaccine Demonstrates Efficacy In Guinea Pigs As Both A Prophylactic And A Therapeutic Vaccine". Retrieved 2 August 2016
- Kolb AW, Ané C, Brandt CR (2013). "Using HSV-1 genome phylogenetics to track past human migrations". PLOS ONE. 8 (10): e76267. doi:10.1371/journal.pone.0076267. PMC 3797750 . PMID 24146849

2

Classification of Gene Therapy

Gene therapy can be categorised as somatic and germline gene therapy. In somatic gene therapy, the curing genes affect only the patient and cannot be passed onto the offspring. In germline gene therapy, the germ cells are modified and therefore it becomes heritable. The major categories of gene therapy are dealt with great details in the section.

Somatic Cell

A somatic or vegetal cell is any biological cell forming the body of an organism; that is, in a multicellular organism, any cell other than a gamete, germ cell, gametocyte or undifferentiated stem cell.

In contrast, gametes are cells that fuse during sexual reproduction, germ cells are cells that give rise to gametes, and stem cells are cells that can divide through mitosis and differentiate into diverse specialized cell types. For example, in mammals, somatic cells make up all the internal organs, skin, bones, blood and connective tissue, while mammalian germ cells give rise to spermatozoa and ova which fuse during fertilization to produce a cell called a zygote, which divides and differentiates into the cells of an embryo. There are approximately 220 types of somatic cells in the human body.

Theoretically, these cells are not germ cells (the source of gametes), they never transmit to their descendants the mutations they have undergone. However, in sponges, non-differentiated somatic cells form the germ line and, in Cnidaria, differentiated somatic cells are the source of the germline.

The word "somatic" is derived from the Greek word *sōma*, meaning "body".

Evolution

As multicellularity evolved many times, sterile somatic cells did too. The evolution of an immortal germline producing specialized somatic cells involved the emergence of mortality, and can be viewed in its simplest version in volvocine algae. Those species with a separation between sterile somatic cells and a germ line are called Weismannists. However, Weismannist development is relatively rare (e.g., vertebrates, arthropods, *Volvox*), as a great number of species have the capacity for somatic embryogenesis (e.g., land plants, most algae, many invertebrates).

Genetics and Chromosome Content

Like all cells, somatic cells contain DNA arranged in chromosomes. If a somatic cell contains chromosomes arranged in pairs, it is called diploid and the organism is called a diploid organism. (The gametes of diploid organisms contain only single unpaired chromosomes and are called haploid.) Each pair of chromosomes comprises one chromosome inherited from the father and one inherited from the mother. For example, in humans, somatic cells contain 46 chromosomes organized into 23 pairs. By contrast, gametes of diploid organisms contain only half as many chromosomes. In humans, this is 23 unpaired chromosomes. When two gametes (i.e. a spermatozoon and an ovum) meet during conception, they fuse together, creating a zygote. Due to the fusion of the two gametes, a human zygote contains 46 chromosomes (i.e. 23 pairs).

However, a large number of species have the chromosomes in their somatic cells arranged in fours ("tetraploid") or even sixes ("hexaploid"). Thus, they can have diploid or even triploid germline cells. An example of this is the modern cultivated species of wheat, *Triticum aestivum L.*, a hexaploid species whose somatic cells contain six copies of every chromatid.

Cloning

In recent years, the technique of cloning whole organisms has been developed in mammals, allowing almost identical genetic clones of an animal to be produced. One method of doing this is called "somatic cell nuclear transfer" and involves removing the nucleus from a somatic cell, usually a skin cell. This nucleus contains all of the genetic information needed to produce the organism it was removed from. This nucleus is then injected into an ovum of the same species which has had its own genetic material removed. The ovum now no longer needs to be fertilized, because it contains the correct amount of genetic material (a diploid number of chromosomes). In theory, the ovum can be implanted into the uterus of a same-species animal and allowed to develop. The resulting animal will be a nearly genetically identical clone to the animal from which the nucleus was taken. The only difference is caused by any mitochondrial DNA that is retained in the ovum, which is different from the cell that donated the nucleus. In practice, this technique has so far been problematic, although there have been a few high-profile successes, such as Dolly the Sheep and, more recently, Snuppy, the first cloned dog. Somatic cells have also been collected in the practice of cryoconservation of animal genetic resources as a means of conserving animal genetic material, including to clone livestock.

Modifications

Development of biotechnology has allowed for the genetic manipulation of somatic cells. This biotechnology deals with some ethical controversy in human genetic engineering.

Somatic Cell Gene Therapy

Somatic cells are all the cells present in the body except the germ cells (oocytes and spermatocytes). Somatic gene therapy is not heritable because the new DNA is inserted into a somatic cell of the body like liver cell and thus the DNA does not get into the germ cells. This transfer corrects the mutation or disorder in an individual suffering from any genetic ailment.This method used for the gene transfer can be either virus mediated or liposome mediated. Sometimes the DNA inserted gets integrated into the chromosome in the nucleus. The cells should be easy to isolate and re-implant like bone marrow cells because it continues to divide for the entire life of the individual producing blood cells. So the diseases having role in blood can be cured this way. Similarly the gene can be inserted wherever its target organ or tissue is located eg- lungs, liver, muscle.

There are certain problems with somatic gene therapy such as risk that the viral vector will combine with the cellular genome and thus infect the cell. This is called as viral escape. Another risk is activation of a proto-oncogene or disruption of an important gene. There are certain hurdles to overcome in somatic cell gene therapy. One is to target the desired cell or tissue to introduce the gene. Virus can be risky at times. Another one is to locate the right cell for example for cystic fibrosis;target should be the lungs and the gut. Aerosol sprays can be used for lungs to deliver the gene and for the gut, the gene to be delivered should be packaged such that it would swallow and still protect them from digestive enzymes until they function.

Somatic cell gene therapies comprise the implantation of cells as an *in vivo* source of any essential biomolecules such as an enzyme, cytokine or coagulation factor, transduction of activated lymphocytes, natural killer cells, and cell populations such as hepatocytes or myocytes which are involved in performing complex biological functions. Initial approach of the gene therapy was to implant the altered cell population in the patient which later on is modified by injecting retroviral vector expressing corrected gene. The complexicities of the therapy includes interaction of cells with other body constituents making the whole science to be tested to utmost care before using it for preclinical trial.

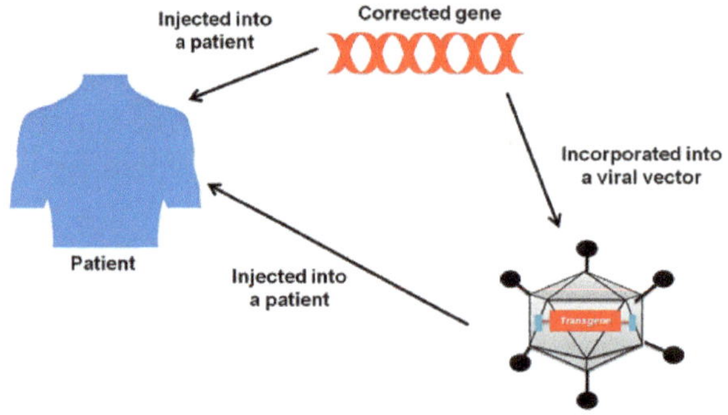

Schematic representation of somatic cell gene therapy:

Target Site For Gene Therapy

There are different target sites for the gene therapy

Endothelium: The advantage of using endothelium as a target site for gene therapy includes formation of new blood vessels and ease to get the gene of interest directly into the blood stream. Eg- Hemophilia

Muscle: It is easily accessible to blood stream and muscle specific promoter is well studied. Eg- Duchenne Muscular Dystrophy.

Liver: It has a capacity to regenerate and involve in a variety of functions. Thyroid binding promoter is well studied for liver specific gene transfer. Eg- Familial hypercholesterolemia.

Skin: Grafting of skin is possible and small piece can be grown to a large.

Brain: Important for nervous system related illnesses.

Lung: Airway epithelium is easily accessible organ. Eg- gene therapy for cystic fibrosis.

Transcriptional Targeting Approach

The specific cells or tissue can be transduced by a vector designed by using specific promoter that directs the expression of gene of interest. The gene expression of these vectors is only possible when the transcription factor proteins can bind to these promoters. Some of the promoter used for expressing the transgene is prostate specific antigen, cytomegalovirus promoter, β-actin promoter, and hypoxia inducible factors.

Germ Cell

A germ cell is any biological cell that gives rise to the gametes of an organism that reproduces sexually. In many animals, the germ cells originate in the primitive streak and migrate via the gut of an embryo to the developing gonads. There, they undergo meiosis, followed by cellular differentiation into mature gametes, either eggs or sperm. Unlike animals, plants do not have germ cells designated in early development. Instead, germ cells can arise from somatic cells in the adult (such as the floral meristem of flowering plants).

Introduction

Multicellular eukaryotes are made of two fundamental cell types. Germ cells produce gametes and are the only cells that can undergo meiosis as well as mitosis. These cells are sometimes said to be immortal because they are the link between generations. Somatic cells are all the other cells that form the building blocks of the body and they only divide by mitosis. The lineage of germ cells is called germ line. Germ cell specification begins during cleavage in many animals or in the epiblast during gastrulation in birds and mammals. After transport, involving passive movements and active migration, germ cells arrive at the developing gonads. In humans, sexual differentiation starts approximately 6 weeks after conception. The end-products of the germ cell cycle are the egg or sperm.

Under special conditions *in vitro* germ cells can acquire properties similar to those of embryonic stem cells (ES). The underlying mechanism of that change is still unknown. These changed cells are then called embryonic germ cells (EG). Both EG and ES are pluripotent in vitro, but only ES has proven pluripotency in vivo. Recent studies have demonstrated that it is possible to give rise to primordial germ cells from ES.

Specification

There are two mechanisms to establish the germ cell lineage in the embryo. The first way is called preformistic and involves that the cells destined to become germ cells inherit the specific germ cell determinants present in the germ plasm (specific area of the cytoplasm) of the egg (ovum). The unfertilized egg of most animals is asymmetrical: different regions of the cytoplasm contain different amounts of mRNA and proteins. By this, germ cells obtained by the first divisions of the fertilized egg are characterized by specific molecules of a particular region of the egg cytoplasm.

The second way is found in birds and mammals, where germ cells are not specified by such determinants but by signals controlled by zygotic genes. In mammals, a few cells of the early embryo are induced by signals of neighboring cells to become primordial germ cells. Mammalian eggs are somewhat symmetrical and after the first divisions of the fertilized egg, the produced cells are all totipotent. This means that they can differentiate in any cell type in the body and thus germ cells. Specification of primordial germ cells in the laboratory mouse is initiated by high levels of bone morphogenetic protein (BMP) signaling, which activates expression of the transcription factors Blimp-1/Prdm1 and Prdm14.

Migration

Primordial germ cells, germ cells that still have to reach the gonads, also known as PGCs, precursor germ cells or gonocytes, divide repeatedly on their migratory route through the gut and into the developing gonads.

Invertebrates

In the model organism *Drosophila*, pole cells passively move from the posterior end of the embryo to the posterior midgut because of the infolding of the blastoderm. Then they actively move through the gut into the mesoderm. Endodermal cells differentiate and together with Wunen proteins they induce the migration through the gut. Wunen proteins are chemorepellents that lead the germ cells away from the endoderm and into the mesoderm. After splitting into two populations, the germ cells continue migrating laterally and in parallel until they reach the gonads. Columbus proteins, chemoattractants, stimulate the migration in the gonadal mesoderm.

Vertebrates

In the *Xenopus* egg, the germ cell determinants are found in the most vegetal blastomeres. These presumptive PGCs are brought to the endoderm of the blastocoel by gastrulation. They are determined as germ cells when gastrulation is completed. Migration from the hindgut along the gut and across the dorsal mesentery then takes place. The germ cells split into two populations and move to the paired gonadal ridges. Migration starts with 3-4 cells that undergo three rounds of cell division so that about 30 PGCs arrive at the gonads. On the migratory path of the PGCs, the orientation of underlying cells and their secreted molecules such as fibronectin play an important role.

Mammals have a migratory path comparable to that in *Xenopus*. Migration begins with 50 gonocytes and about 5,000 PGCs arrive at the gonads. Proliferation occurs also during migration and lasts for 3–4 weeks in humans.

PGCs come from the epiblast and migrate subsequently into the mesoderm, the endoderm and the posterior of the yolk sac. Migration then takes place from the hindgut along the gut and across the dorsal mesentery to reach the gonads (4.5 weeks in human beings). Fibronectin maps here also a polarized network together with other molecules. The somatic cells on the path of germ cells provide them attractive, repulsive, and survival signals. But germ cells also send signals to each other.

In reptiles and birds, germ cells use another path. PGCs come from the epiblast and move to the hypoblast to form the germinal crescent (anterior extraembryonic structure). The gonocytes then squeeze into blood vessels and use the circulatory system for transport. They squeeze out of the vessels when they are at height of the gonadal ridges. Cell adhesion on the endothelium of the blood vessels and molecules such as chemoattractants are probably involved in helping PGCs migrate.

The *Sry* Gene of the Y Chromosome

The *SRY* (**S**ex-determining **R**egion of the **Y** chromosome) directs male development in mammals by inducing the somatic cells of the gonadal ridge to develop into a testis, rather than an ovary. *Sry* is expressed in a small group of somatic cells of the gonads and influences these cells to become Sertoli cells (supporting cells in testis). Sertoli cells are responsible for sexual development along a male pathway in many ways. One of these ways involves stimulation of the arriving primordial cells to differentiate into sperm. In the absence of the *Sry* gene, primordial germ cells differentiate into eggs. Removing genital ridges before they start to develop into testes or ovaries results in the development of a female, independent of the carried sex chromosome.

Gametogenesis

Gametogenesis, the development of diploid germ cells into either haploid eggs or sperm, (respectively oogenesis and spermatogenesis) is different for each species but the general stages are similar. Oogenesis and spermatogenesis have many features in common, they both involve:

- Meiosis
- Extensive morphological differentiation
- Incapacity of surviving for very long if fertilization does not occur

Despite their homologies they also have major differences:

- Spermatogenesis has equivalent meiotic divisions resulting in four equivalent spermatids while oogenic meiosis is asymmetrical: only one egg is formed together with three polar bodies.
- Different timing of maturation: oogenic meiosis is interrupted at one or more stages (for a long time) while spermatogenic meiosis is rapid and uninterrupted.

Oogenesis

After migration primordial germ cells will become oogonia in the forming gonad (ovary). The oogonia proliferate extensively by mitotic divisions, up to 5-7 million cells in humans. But then many of these oogonia die and about 50,000 remain. These cells differentiate into primary oocytes. In week 11-12 *post coitus* the first meiotic division begins (before birth for most mammals) and remains arrested in prophase I from a few days to many years depending on the species. It is in this period or in some cases at the beginning of sexual maturity that the primary oocytes secrete proteins to form a coat called zona pellucida and they also produce cortical granules containing enzymes and proteins needed for fertilization. Meiosis stands by because of the follicular granulosa cells that send inhibitory signals through gap junctions and the zona pellucida. Sexual maturation is the beginning of periodic ovulation. Ovulation is the regular release of one oocyte from the ovary into the reproductive tract and is preceded by follicular growth. A few follicle cells are stimulated to grow but only one oocyte is ovulated. A primordial follicle consists of an epithelial layer of follicular granulosa cells enclosing an oocyte. The pituitary gland secrete follicle-stimulating hormones (FSHs) that stimulate follicular growth and oocyte maturation. The thecal cells around each follicle secrete estrogen. This hormone stimulates the production of FSH receptors on the follicular granulosa cells and has at the same time a negative feedback on

FSH secretion. This results in a competition between the follicles and only the follicle with the most FSH receptors survives and is ovulated. Meiotic division I goes on in the ovulated oocyte stimulated by luteinizing hormones (LHs) produced by the pituitary gland. FSH and LH block the gap junctions between follicle cells and the oocyte therefore inhibiting communication between them. Most follicular granulosa cells stay around the oocyte and so form the cumulus layer. Large non-mammalian oocytes accumulate egg yolk, glycogen, lipids, ribosomes, and the mRNA needed for protein synthesis during early embryonic growth. These intensive RNA biosynthese are mirrored in the structure of the chromosomes, which decondense and form lateral loops giving them a lampbrush appearance. Oocyte maturation is the following phase of oocyte development. It occurs at sexual maturity when hormones stimulate the oocyte to complete meiotic division I. The meiotic division I produces 2 cells differing in size: a small polar body and a large secondary oocyte. The secondary oocyte undergoes meiotic division II and that results in the formation of a second small polar body and a large mature egg, both being haploid cells. The polar bodies degenerate. Oocyte maturation stands by at metaphase II in most vertebrates. During ovulation, the arrested secondary oocyte leaves the ovary and matures rapidly into an egg ready for fertilization. Fertilization will cause the egg to complete meiosis II. In human females there is proliferation of the oogonia in the fetus, meiosis starts then before birth and stands by at meiotic division I up to 50 years, ovulation begins at puberty.

Egg Growth

A 10 - 20 μm large somatic cell generally needs 24 hours to double its mass for mitosis. By this way it would take a very long time for that cell to reach the size of a mammalian egg with a diameter of 100 μm (some insects have eggs of about 1,000 μm or greater). Eggs have therefore special mechanisms to grow to their large size. One of these mechanisms is to have extra copies of genes: meiotic division I is paused so that the oocyte grows while it contains two diploid chromosome sets. Some species produce many extra copies of genes, such as amphibians, which may have up to 1 or 2 million copies. A complementary mechanism is partly dependent on syntheses of other cells. In amphibians, birds, and insects, yolk is made by the liver (or its equivalent) and secreted into the blood. Neighboring accessory cells in the ovary can also provide nutritive help of two types. In some invertebrates some oogonia become nurse cells. These cells are connected by cytoplasmic bridges with oocytes. The nurse cells of insects provide oocytes macromolecules such as proteins and mRNA. Follicular granulosa cells are the second type of accessory cells in the ovary in both invertebrates and vertebrates. They form a layer around the oocyte and nourish them with small molecules, no macromolecules, but eventually their smaller precursor molecules, by gap junctions.

Spermatogenesis

Mammalian spermatogenesis is representative for most animals. In human males, spermatogenesis begins at puberty in seminiferous tubules in the testicles and go on continuously. Spermatogonia are immature germ cells. They proliferate continuously by mitotic divisions around the outer edge of the seminiferous tubules, next to the basal lamina. Some of these cells stop proliferation and differentiate into primary spermatocytes. After they proceed through the first meiotic division, two secondary spermatocytes are produced. The two secondary spermatocytes undergo the second meiotic division to form four haploid spermatids. These spermatids differentiate morphologically into sperm by nuclear condensation, ejection of the cytoplasm and formation of the acrosome and flagellum.

The developing male germ cells do not complete cytokinesis during spermatogenesis. Consequently, cytoplasmic bridges assure connection between the clones of differentiating daughter cells to form a syncytium. In this way the haploid cells are supplied with all the products of a complete diploid genome. Sperm that carry a Y chromosome, for example, is supplied with essential molecules that are encoded by genes on the X chromosome.

Diseases

Germ cell tumor is a rare cancer that can affect people at all ages. 2.4 children out of 1 million are affected, and it counts for 4% of all cancers in children and adolescents younger than 20 years old.

Germ cell tumors are generally located in the gonads but can also appear in the abdomen, pelvis, mediastinum, or brain. Germ cells migrating to the gonads may not reach that intended destination and a tumor can grow wherever they end up, but the exact cause is still unknown. These tumors can be benign or malignant.

Induced Differentiation

Inducing differentiation of certain cells to germ cells has many applications. One implication of induced differentiation is that it may allow for the eradication of male and female factor infertility. Furthermore, it would allow same-sex couples to have biological children if sperm could be produced from female cells or if eggs could be produced from male cells. Efforts to create sperm and eggs from skin and embryonic stem cells were pioneered by Hayashi and Saitou's research group at Kyoto University. These researchers produced primordial germ cell-like cells (PGCs) from embryonic stem cells (ESCs) and skin cells in vitro.

Hayashi and Saitou's group was able to promote the differentiation of embryonic stem cells into PGCs with the use of precise timing and bone morphogenetic protein 4 (Bmp4). The PGCs were then placed in the testes of mice that were initially unable to produce their own sperm cells. As a result, the mice were able to produce sperm cells. Upon succeeding with embryonic stem cells, the group was able to successfully promote the differentiation of induced pluripotent stem cells (iPSCs) into PGCs. These primordial germ cell-like cells were then used to create spermatozoa and oocytes.

Efforts for human cells are less advanced due to the fact that the PGCs formed by these experiments are not always viable. In fact Hayashi and Saitou's method is only one third as effective as current in vitro fertilization methods, and the produced PGCs are not always functional. Furthermore, not only are the induced PGCs not as effective as naturally occurring PGCs, but they are also less effective at erasing their epigenetic markers when they differentiate from iPSCs or ESCs to PGCs.

There are also other applications of induced differentiation of germ cells. Another study showed that culture of human embryonic stem cells in mitotically inactivated porcine ovarian fibroblasts (POF) causes differentiation into germ cells, as evidenced by gene expression analysis.

Germ Line Gene Therapy

The germ cells (sperm and egg) are manipulated in germ line gene therapy. The foreign DNA is added to the germ cells or early embryo. The advantage is that they are available outside the body easily and can thus be retrieved and modified and the inserted gene would be present in all the cells of the patient and would also be passed on to further generations. Thus it will be heritable. It is good for the people with inheritable genetic disorders. But the problem is targeting the gene to the appropriate location as it may get inserted to some other location in the genome of the cell thus causing mutation or disrupting a functional normal gene.

In germline gene therapy desirable changes or modification in the targeted genes may be done in two ways. In the first type modification of the genetic constitution of germ cells is done prior to the fertilization and hence the modifications in the genetic constitution pass on to the later generations in the form of mutation. However in present scenario this therapy is considered undesirable because of the huge risk associated with it.

In the second type of germline gene therapy changes in the genetic constitution are brought in the initial stages of blastomere. In this type the altered cells are allowed to divide and grow into an individual and only few of these cells acquire the genetic changes and remaining do not. The mutation will be inherited only if the cells acquiring the genetic changes turn into germ cells of the organism.

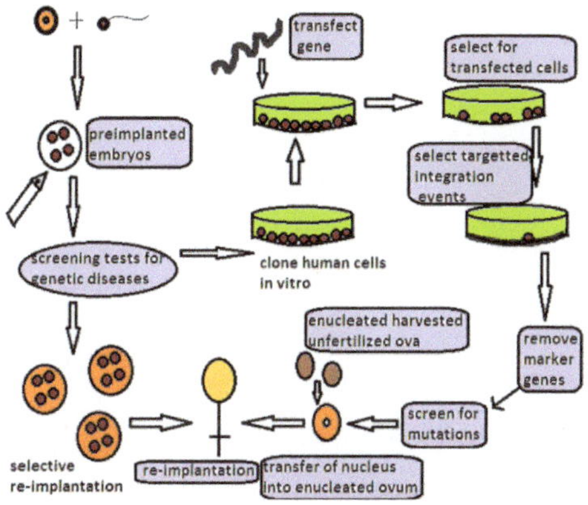

Schematic representation of germline gene therapy

Germ line gene therapy has just started although experiments have been carried out in animals and some mammals. This therapy is still unacceptable ethically as no trials have been conducted on humans till date.

Steps of Germ Line Gene Therapy:

1. Isolation of totipotent embryonic cells at an undifferentiated stage
2. The determination of the genetic state of the embryo
3. Embryonic stem cell expansion in culture
4. Transfer of the genetic material into embryonic cell
5. Selection of stably transfected cell
6. Targeted gene replacement
7. Removal of marker
8. Confirmation of genomic integrity
9. Transfer of the nucleus
10. Re-implantation in the mother

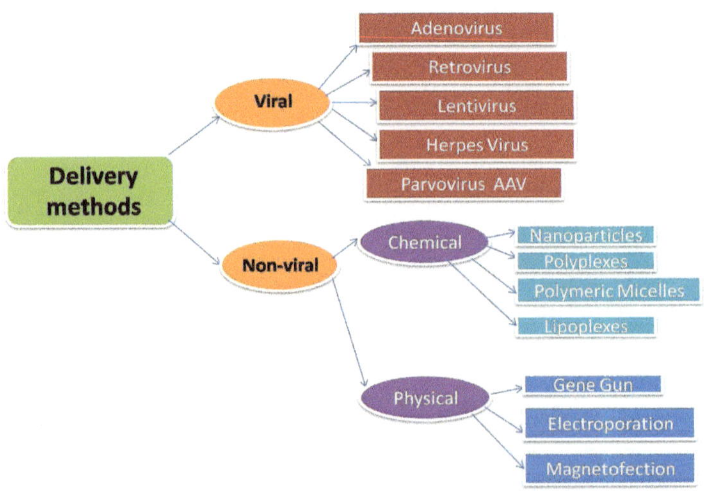

Different delivery methods

Gene Replacement

The techniques of genome mapping, mutation, protein expression have played a great role in understanding modern molecular cell biology. However, large number of genes have been recognized and studied on the basis of the biochemical properties of their encoded protein, the sequence similarity of the encoded protein with proteins of established role, or their policy of expression during their progression. It would be very difficult to understand the *in vivo* role of such genes if their mutant forms are not available. Gene replacement is a technique wherein a particular gene is mutated *in vitro* and its native copy is then substituted by its mutant form to know its *in vivo* function. Gene replacement is also sometimes known as gene-targeted knockout , or plainly " knockout" . In other words gene replacement is a technique that involves homologous recombination to replace or substitute an endogenous gene. It may include point mutations, removal of a gene/exons and even introduction of a new gene. Gene replacement or knockdown can be constant or provisional. Provisional or limited knockdown can be in terms of its restriction to a particular period during the growth of an organism or restriction to a particular tissue.

Gene replacement technique is an established method for some typical organisms and the method may vary among the wide range of species used. Overall, a construct made out of DNA is produced in bacteria. It generally comprises of targeted gene, a reporter gene and a selectable marker that helps in suitable selection. *Drosophila melanogaster* is the species of choice for many genetic experiments and in likewise species gene therapy has shown its prompt role. In some cases especially knockout experiments exhibit high frequency by using engineered endonucleases such as Zinc finger nucleases (ZFN) and homing endonucleases. In knockout mice experiments the above mentioned construct is inserted into the mouse embryonic stem cells in culture. The cells with proper insertion are then picked in order to use them for contributing to a mouse's tissue through embryo injection. Lastly, only the chimeric mice in which the reproductive organs are made up of modified cells are taken and selected for breeding. This step ensures that the complete body of the mouse is based on the originally picked embryonic stem cell. In present scenario, gene knockout has now become possible in many other organisms including yeast, cattle, sheep, swine, and many fungi.

With the ease in learning the skills of genome sequencing of various model organisms mutating a native gene would probably become quite straightforward. In any case (native protein or sequenced genome) the process can be compiled in figure.

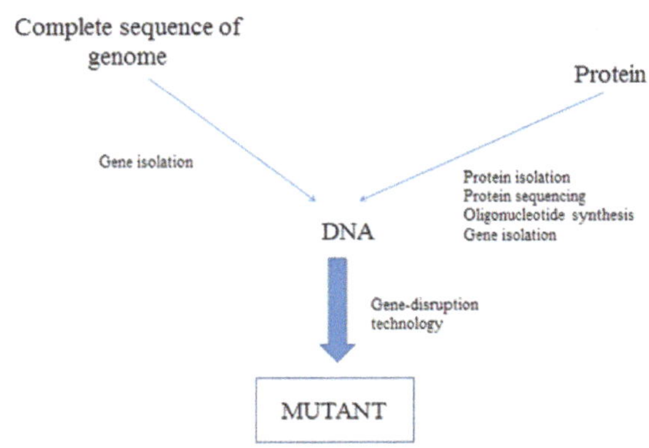

Complete genome of an organism and its application

The other technique which involves integration of exogenous gene called a transgene into the host organism to display a new function in that particular organism is called as transgenesis. The assembly of both gene replacement and transgenesis makes use of methods for modifying cloned genes *in vitro* and then relocate them into eukaryotic cells.

One of the successful applications of gene replacement technology is the experimentation carried out in genetic engineering to eliminate unacceptable metabolic pathways in *Streptomyces clavuligerus* which is an industrial strain used for the synthesis of an antibiotic significant for the human medicine.

Important Terminology:

Chimera – It is also read as chimaera and may be defined as an organism that comprises of two or more than two varied populations of genetically distinct cells that originate from various zygotes participating in sexual reproduction. If the varying cells arise from the same zygote, the organism is termed a mosaic .

Gene Addition

In present scenario gene addition is one of the most practical approaches to gene therapy. It involves the insertion of an active copy of a defective innate gene. In the year 1990, viral vector based gene addition was successfully done to treat a patient named Thompson in the U.S. In the viral vector based gene addition Retroviruses were used at first because they were supposed and found to be most fit among all viruses for their adaptive nature for delivering genes into cells. This therapy is a promising one because it has been proven to be a substitute for gene replacement and is having a very high efficiency for curing any disease. Even though gene addition therapy is quite satisfactory but it has some benefits and drawbacks over the gene replacement therapy. Its advantages are that the cells that do not ordinarily express a specific gene can be manipulated to express that gene. For example, blood clotting factors could be synthesized in any somatic cell, making somatic tissues other than the liver convenient objectives for clotting factor gene transfer. Alternately, completely new genes might be used to cure the disease. For example, the transmission of human immunodeficiency virus (HIV) could be restricted by transfer of genes that encode ribozymes competent of degrading HIV RNA. A disadvantage of gene addition therapy is the disordered insertion of genes into the genome. Thus, the inserted genes may be erroneously expressed or may cause inappropriate expression of genes immediate to the insertion site.

The outcome of gene addition relies on the effective insertion of therapeutic genes at the suitable target spot within the host cell, avoiding any cell damage, mutations or any kind of adverse immune response . Zinc finger nucleases (ZFNs) have become essential reagents for engineering the genomes of many plants and animals including *drosophila melano-*

gaster . Specially constructed laboratory, ZFNs suggests a common method to deliver a site-specific double strand breaks (DSB) to the genome, and stimulate a local homologous recombination by considerable amount of magnitude. The ZFN-encoding plasmid-based approach has the efficacy to avoid all the problems linked with the viral delivery of therapeutic genes. ZFNs drive exceedingly effective genome editing by generating a site-specific DNA DSB at a fixed site in the genome. Consequent restoration of this DSB break through the homology-directed repair (HDR) leads to targeted gene addition.

In some wet lab experiments addition of a break at a specific target site causes gene addition using a homologous donor template on using CCR5-specific ZFNs as a model system. Efficacy for a single stranded break to direct repair pathway choice may prove beneficial for some therapeutic applications such as the targeted correction of human disease-causing mutations.

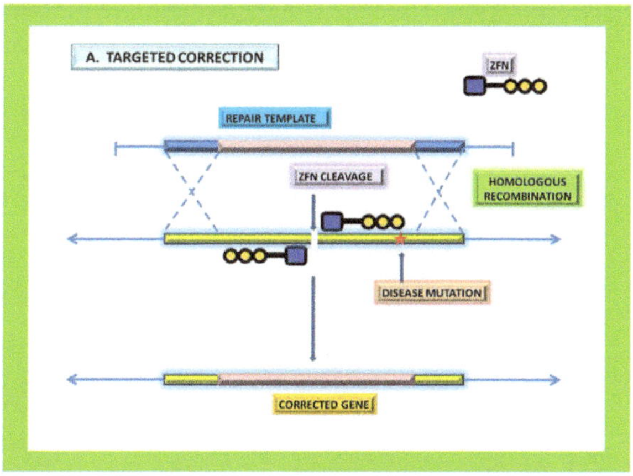

Diagrammatic illustration of targeted gene correction strategy

In the figure expressed ZFNs pairs bind to and break the target genomic locus close to the disease-causing mutation. ZFNs comprise of chimeric zinc fingers (yellow) and *Fok1* endonuclease domains (blue). Further the repair template donor is used for the insertion of a DSB that stimulates homologous recombination.

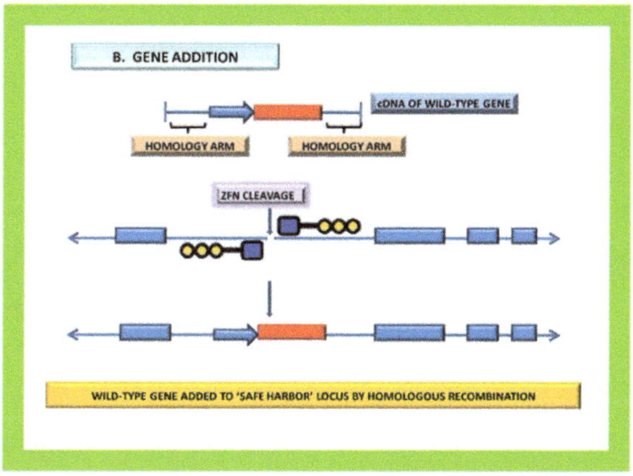

Diagrammatic illustration of targeted gene addition strategy

In the figure that discusses targeted gene addition, the ZFN pair cleaves a predetermined 'safe harbor' locus. The expression cassette with a therapeutic gene (red) is bounded by homology arms to the safe harbor locus.

Gene Doping

Gene doping is the hypothetical non-therapeutic use of gene therapy in order to improve athletic performance in competitive sporting events. As of April 2015, there is no evidence that gene doping has been used for athletic performance-enhancement in any sporting events. Gene doping would involve the use of gene transfer to increase or decrease gene expression and protein biosynthesis of a specific human protein; this could be done by directly injecting the gene carrier into the person, or by taking cells from the person, transfecting the cells, and administering the cells back to the person.

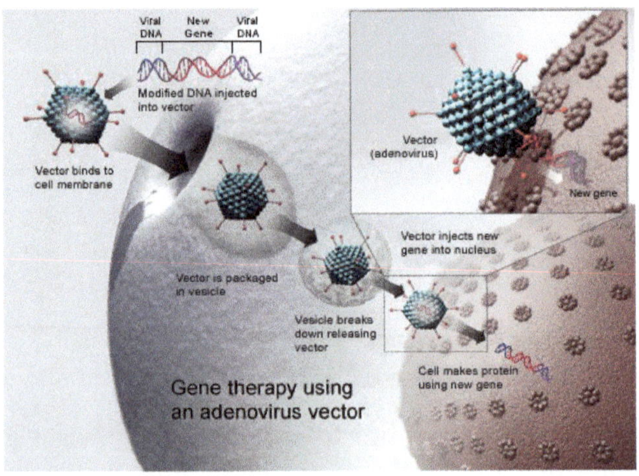

An illustration of viral vector-mediated gene transfer using an adenovirus as the vector.

The historical development of interest in gene doping by athletes and concern about the risks of gene doping and how to detect it moved in parallel with the development of the field of gene therapy, especially with the publication of work on a transgenic mouse overexpressing insulin-like growth factor 1 that was much stronger than normal mice, even in old age, preclinical studies published in a way to deliver erythropoietin (EPO) via gene therapy, and publication in the creation of a "marathon mouse" with much greater endurance than normal mice, created by delivering the gene expressing PPAR gamma to the mice. The scientists generating these publications were all contacted directly by athletes and coaches seeking access to the technology. The public became aware of that activity in 2006 when such efforts were part of the evidence presented in the trial of a German coach.

Scientists themselves, as well as bodies including the World Anti-Doping Agency (WADA), the International Olympic Committee, and the American Association for the

Advancement of Science, started discussing the risk of gene doping in 2001, and by 2003 WADA had added gene doping to the list of banned doping practices, and shortly thereafter began funding research on methods to detect gene doping.

History

The history of concern about the potential for gene doping follows the history of gene therapy, the medical use of genes to treat diseases, which was first clinically tested in the 1990s. Interest by the athletic community was especially spurred by the creation in a university lab of a "mighty mouse", created by administering a virus carrying the gene expressing insulin-like growth factor 1 to mice; the mice were stronger and remained strong even as they aged, without exercise. The lab had been seeking treatments for muscle wasting diseases, but when their work was made public, the lab was inundated with calls from athletes seeking the treatment, with one coach offering his whole team. The scientist told *The New York Times* in 2007: "I was quite surprised, I must admit. People would try to entice me, saying things like, 'It'll help advance your research.' Some offered to pay me." He also told the *Times* that every time similar research is published he gets calls and that he explains that, even should the treatment became ready for use in people, which would take years, there would be serious risks, including death; he also said that even after he explains this, the athletes still want it.

In 1999, the field of gene therapy was set back when Jesse Gelsinger died in a gene therapy clinical trial, suffering a massive inflammatory reaction to the drug. This led regulatory authorities in the US and Europe to increase safety requirements in clinical trials even beyond the initial restrictions that had been put in place at the beginning of the biotechnology era to deal with the risks of recombinant DNA.

In June 2001, Theodore Friedmann, one of the pioneers of gene therapy, and Johann Olav Koss an Olympic gold medallist in speed skating, published a paper that was the first public warning about gene doping. Also in June 2001, a Gene Therapy Working Group, convened by the Medical Commission of the International Olympic Committee noted that "we are aware that there is the potential for abuse of gene therapy medicines and we shall begin to establish procedures and stateof-the-art testing methods for identifying athletes who might misuse such technology."

Research was published in 2002 about a preclinical gene therapy called Repoxygen, which delivered the gene encoding erythropoietin (EPO) as a potential treatment for anemia. The scientists from that company also received calls from athletes and coaches. In that same year the World Anti-Doping Agency held its first meeting to discuss the risk of gene doping, and the US President's Council on Bioethics discussed gene doping in the context of human enhancement at several sessions.

In 2003, the field of gene therapy took a step forward and a step back; first gene therapy drug was approved, Gendicine, which was approved in China for the treatment of certain cancers, but children in France who had seemingly been effective treated with gene therapy for Severe combined immunodeficiency (non-human) began developing leukemia. In 2003 the BALCO scandal became public, in which chemists, trainers and athletes conspired to evade doping controls with new and undetectable doping substances. In 2003 the World Doping Agency proactively added gene doping to the list of banned doping practices. Also in 2003, a symposium convened by the American Association for the Advancement of Science focused on the issue.

Research published in 2004 showing that mice given gene therapy coding for a protein called PPAR gamma had about double the endurance of untreated mice and were dubbed "marathon mice"; those scientists received calls from athletes and coaches. Also in 2004 the World Anti-Doping Agency began to fund research to detect gene doping, and formed a permanent expert panel to advise it on risks and to guide the funding.

In 2006 interest from athletes in gene doping received widespread media coverage due its mention during the trial of a German coach who was accused and found guilty of giving his athletes performance enhancing drugs without their knowledge; an email in which the coach attempted to obtain Repoxygen was read in open court by a prosecutor. This was the first public disclosure that athletes were interested in gene doping.

In 2011 the second gene therapy drug was approved; Neovasculgen, which delivers the gene encoding VEGF, was approved in Russia to treat peripheral artery disease.

In 2012 Glybera, a treatment for a rare inherited disorder, became the first treatment to be approved for clinical use in either Europe or the United States.

As the field of gene therapy has developed, the risk of gene doping becoming a reality has increased with it.

Agents

There are numerous genes of interest as agents for gene doping. They include erythropoietin, insulin-like growth factor 1, human growth hormone, myostatin, vascular endothelial growth factor, fibroblast growth factor, endorphin, enkephalin and Alpha-actinin-3.

The risks of gene doping would be similar to those of gene therapy: immune reaction to the native protein leading to the equivalent of an genetic disease, massive inflammatory response, cancer, and death, and in all cases, these risks would be undertaken for short-term gain as opposed to treating a serious disease.

Alpha-actinin-3

Alpha-actinin-3 is found only in skeletal muscle in humans, and has been identified in several genetic studies as having a different polymorphism in world-class athletes compared with normal people. One form that causes the gene to make more protein is found in sprinters and is related to increased power; another form that causes the gene to make less protein is found in endurance athletes. Gene doping agents could be designed with either polymorphism, or for endurance athletes, some DNA construct that interfered with expression like a small interfering RNA.

Myostatin

Myostatin is a protein responsible for inhibiting muscle differentiation and growth. Removing the myostatin gene or otherwise limiting its expression leads to an increase in muscle size and power. This has been demonstrated in knockout mice lacking the gene that were dubbed "Schwarzenegger mice". Humans born with defective genes can also serve as "knockout models"; a German boy with a mutation in both copies of the myostatin gene was born with well-developed muscles. The advanced muscle growth continued after birth, and the boy could lift weights of 3 kg at the age of 4. In work published in 2009, scientists administered follistatin via gene therapy to the quadriceps of non-human primates, resulting in local muscle growth similar to the mice.

Erythropoietin (EPO)

Erythropoietin is a glycoprotein that acts as a hormone, controlling red blood cell production. Athletes have injected the EPO protein as a performance-enhancing substance for many years (blood doping). When the additional EPO increases the production of red blood cells in circulation, this increases the amount of oxygen available to muscle, enhancing an athlete's endurance. Recent studies suggest it may be possible to introduce another EPO gene into an animal in order to increase EPO production endogenously. EPO genes have been successfully inserted into mice and monkeys, and were found to increase hematocrits by as much as 80 percent in those animals. However, the endogonous and transgene derived EPO elicited autoimmune responses in some animals in the form of severe anemia.

Insulin-like Growth Factor 1

Insulin-like growth factor 1 is a protein involved in the mediation of the growth hormone. Administration of IGF-1 to mice has resulted in more muscle growth and quicker muscle and nerve regeneration. If athletes were to use this the sustained production of IGF-1 could cause heart disease and cancer.

Others

Modulating the levels of proteins that affect psychology are also potential goals for gene doping; for example pain perception depends on endorphins and enkephalins, response to stress depends on BDNF, and an increase in synthesis of monamines could improve the mood of athletes. Preproenkephalin has been administered via gene therapy using a replication-deficient herpes simplex virus, which targets nerves, to mice with results good enough to justify a Phase I clinical trial in people with terminal cancer with uncontrolled pain. Adopting that approach for athletes would be problematic since the pain deadening would likely be permanent.

VEGF has been tested in clinical trials to increase blood flow and has been considered as a potential gene doping agent; however long term follow up of the clinical trial subjects showed poor results. The same is true of fibroblast growth factor. Glucagon-like peptide-1 increases the amount of glucose in the liver and has been administered via gene therapy to the livers of mouse models of diabetes and was shown to increase gluconeogenesis' for athletes this would make more energy available and reduce the build-up of lactic acid.

Detection

The World Anti-Doping Agency (WADA) is the main regulatory organization looking into the issue of the detection of gene doping. Both direct and indirect testing methods are being researched by the organization. Directly detecting the use of gene therapy usually requires the discovery of recombinant proteins or gene insertion vectors, while most indirect methods involve examining the athlete in an attempt to detect bodily changes or structural differences between endogenous and recombinant proteins.

Indirect methods are by nature more subjective, as it becomes very difficult to determine which anomalies are proof of gene doping, and which are simply natural, though unusual, biological properties. For example, Eero Mäntyranta, an Olympic cross country skier, had a mutation which made his body produce abnormally high amounts of red blood cells. It would be very difficult to determine whether or not Mäntyranta's red blood cell levels were due to an innate genetic advantage, or an artificial one.

Research

A 2016 review found that about 120 DNA polymorphisms had been identified in the literature related to some aspect of athletic performance, 77 related to endurance and 43 related to power. 11 had been replicated in three or more studies and six were identified in genome-wide association studies, but 29 had not been replicated in at least one study.

The 11 replicated markers were:

Endurance

- ACE Alu I/D (rs4646994) (Called ACE I)
- ACTN3 577X
- PPARA rs4253778 G,
- PPARGC1A Gly482;

power/strength markers

- ACE Alu I/D (rs4646994) (called ACE D)
- ACTN3 Arg577
- AMPD1 Gln12
- HIF1A 582Ser
- MTHFR rs1801131 C
- NOS3 rs2070744 T
- PPARG 12Ala

The six GWAS markers were:

- CREM rs1531550 A,
- DMD rs939787 T
- GALNT13 rs10196189 G
- NFIA-AS1 rs1572312 C,
- RBFOX1 rs7191721 G
- TSHR rs7144481 C

Ethics

The mainstream perspective is that gene doping is dangerous and unethical, as is any application of a therapeutic intervention for non-therapeutic or enhancing purposes, and that it compromises the ethical foundation of medicine and the spirit of sport. Others, who support human enhancement on broader grounds, or who see a false dichotomy between "natural" and "artificial" or a denial of the role of technology in improving athletic performance, do not oppose or support gene doping.

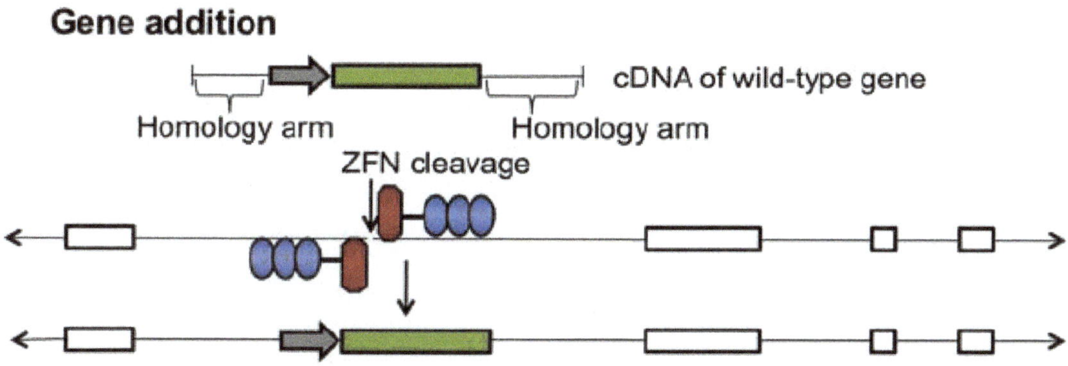

Gene Therapy of the Human Retina

Retinal gene therapy holds great promise in treating different forms of non-inherited and inherited blindness.

In 2008, three independent research groups reported that patients with the rare genetic retinal disease Leber's Congenital Amaurosis had been successfully treated using gene therapy with adeno-associated virus (AAV). In all three studies, an AAV vector was used to deliver a functional copy of the RPE65 gene, which restored vision in children suffering from LCA. These results were widely seen as a success in the gene therapy field, and have generated excitement and momentum for AAV-mediated applications in retinal disease.

In retinal gene therapy, the most widely used vectors for ocular gene delivery are based on adeno-associated virus. The great advantage in using adeno-associated virus for the gene therapy is that it poses minimal immune responses and mediates long-term transgene expression in a variety of retinal cell types. For example, tight junctions that form the blood-retina barrier, separate subretinal space from the blood supply, providing protection from microbes and decreasing most immune-mediated damages.

Clinical Trials

Leber's Congenital Amaurosis

In 2008, three groups reported positive results of clinical trials using adeno-associated virus for Leber's Congenital Amaurosis. In these studies, an AAV vector encoding the RPE65 gene was delivered via a "subretinal injection," where a small amount of fluid is injected underneath the retina in a short surgical procedure.

Initial results from all three studies indicate that AAV is safe in the retina, with no dose-limiting toxicities observed. Across the three trials, no serious adverse events were observed. Furthermore, patients in all three studies showed improvement in their visual function as measured by a number of methods. The methods used varied among the three trials, but included both functional methods such as visual acuity and functional mobility as well as objective measures that are less susceptible to bias, such as the pupil's ability to respond to light and improvements on functional MRI. Improvements were sustained over the long-term, with patients continuing to do well after more than 1.5 years.

UPenn Trial

The UPenn trial was led by a group of four principal investigators at the University of Pennsylvania and the Children's Hospital of Philadelphia: Jean E. Bennett, MD PhD, Albert Maguire MD, Katherine High MD, and J. Fraser Wright, PhD. In this trial, patients improved in terms of visual acuity, pupillometry, visual field, light sensitivity, mobility, and functional MRI. In the Phase 1/2 clinical trial, children as young as 8 years old were treated, and results indicated that younger patients had better results compared to older patients. A company called Spark therapeutics was spun out of the work done at the UPenn in 2013. Spark was granted a "breakthrough-therapy" designation by the FDA and in November 2014 the company was running a phase III trial of their gene therapy product SPK-RPE65. Spark planned to submit results to the FDA in 2016.

Florida Trial

The Florida trial was conducted under the leadership of Dr. Samuel Jacobson and Dr. William Hauswirth, with funding support from the National Eye Institute. This trial treated adults over the age of 18. The treatment was delivered safely, and was associated with improvements in vision. Unlike the UPenn trial, the injection was not delivered underneath the fovea, but next to it. As patients' vision improved, they had the best vision in the injected area, rather than in the usual place at the fovea. This led to the development of a so-called "pseudo-fovea" that corresponded to the treated area.

Age-related Macular Degeneration

Following the successful clinical trials in LCA, researchers have been developing similar treatments using adeno-associated virus for age-related macular degeneration (AMD). To date, efforts have focused on long-term delivery of VEGF inhibitors to treat the wet form of macular degeneration. Whereas wet AMD is currently treated using frequent injections of recombinant protein into the eyeball, these goal of these treatments is long-term disease management following a single administration. One such study is being conducted at the Lions Eye Institute in Australia in collaboration with Avalanche Biotechnologies, a US-based biotechnology start-up. Another early-stage study is sponsored by Genzyme Corporation.

Choroideremia

In October 2011, the first clinical trial was announced for the treatment of choroideremia. Dr. Robert MacLaren of the University of Oxford, who lead the trial, co-developed the treatment with Dr. Miguel Seabra of the Imperial College, London. This Phase 1/2 trial used subretinal AAV to restore the REP gene in affected patients. Initial results of the trial were reported in January 2014 as promising as all six patients had better vision.

Color Blindness

Recent research has shown that AAV can successfully restore color vision to treat color blindness in adult monkeys. Although this treatment has not yet entered clinical trials for humans, this work was considered a breakthrough for the ability to target cone photoreceptors.

Mechanism

Physiological Components in Retinal Gene Therapy

The vertebrate neural retina composed of several layers and distinct cell types. A number of these cell types are implicated in retinal diseases, including retinal ganglion cells, which degenerate in glaucoma, the rod and cone photoreceptors, which are responsive to light and degenerate in retinitis pigmentosa, macular degeneration, and other retinal diseases, and the retinal pigment epithelium (RPE), which supports the photoreceptors and is also implicated in retinitis pigmentosa and macular degeneration.

In retinal gene therapy, AAV is capable of "transducing" these various cell types by entering the cells and expressing the therapeutic DNA sequence. Since the cells of the retina are non-dividing, AAV continues to persist and provide expression of the therapeutic DNA sequence over a long time period that can last several years.

AAV Tropism and Routes of Administration

AAV is capable of transducing multiple cell types within the retina. AAV serotype 2, the most well-studied type of AAV, is commonly administered in one of two routes: intravitreal or subretinal. Using the intravitreal route, AAV is injected in the vitreous humor of the eye. Using the subretinal route, AAV is injected underneath the retina, taking advantage of the potential space between the photoreceptors and RPE layer, in a short surgical procedure. Although this is more invasive than the intravitreal route, the fluid is absorbed by the RPE and the retina flattens in less than 14 hours without complications. Intravitreal AAV targets retinal ganglion cells and a few Muller glial cells. Subretinal AAV efficiently targets photoreceptors and RPE cells.

The reason that different routes of administration lead to different cell types being transfected (e.g., different tropism) is that the inner limiting membrane (ILM) and the various retinal layers act as physical barriers for the delivery of drugs and vectors to the deeper retinal layers. Thus overall, subretinal AAV is 5-10 times more efficient than delivery using the intravitreal route.

Tropism Modification and Novel AAV Vectors

One important factor in gene delivery is developing altered cell tropisms to narrow or broaden rAAV-mediated gene delivery and to increase its efficiency in tissues. Specific properties like capsid conformation, cell targeting strategies can determine which cell types are affected and also the efficiency of the gene transfer process. Different kinds of modification can be undertaken. For example, modification by chemical, immunological or genetic changes that enables the AAV2 capsid to interact with specific cell surface molecules.

Initial studies with AAV in the retina have utilized AAV serotype 2. Researchers are now beginning to develop new variants of AAV, based on naturally-occurring AAV serotypes and engineered AAV variants.

Several naturally-occurring serotypes of AAV have been isolated that can transduce retinal cells. Following intravitreal injection, only AAV serotypes 2 and 8 were capable of transducing retinal ganglion cells. Occasional Muller cells were transduced by AAV serotypes 2, 8, and 9. Following subretinal injection, serotypes 2, 5, 7, and 8 efficiently transduced photoreceptors, and serotypes 1, 2, 5, 7, 8, and 9 efficiently transduce RPE cells.

One example of an engineered variant has recently been described that efficiently transduces Muller glia following intravitreal injection, and has been used to rescue an animal model of aggressive, autosomal-dominant retinitis pigmentosa.

AAV and Immune Privilege in the Retina

Importantly, the retina is immune-privileged, and thus does not experience a significant inflammation or immune-response when AAV is injected. Immune response to gene therapy vectors is what has caused previous attempts at gene therapy to fail, and is considered a key advantage of gene therapy in the eye. Re-administration has been successful in large animals, indicating that no long-lasting immune response is mounted.

Recent data indicates that the subretinal route may be subject to a greater degree of immune privilege compared to the intravitreal route.

Promoter Sequence

Expression in various retinal cell types can be determined the promoter sequence. In order to restrict expression to a specific cell type, a tissue-specific or cell-type specific promoter can be used.

For example, in rat the murine rhodopsin gene drive the expression in AAV2, GFP reporter product was found only in rat photoreceptors, not in any other retinal cell type or in the adjacent RPE after subretinal injection. On the other hand, if ubiquitously expressed immediate-early cytomegalovirus (CMV) enhancer-promoter is expressed in a wide variety of transfected cell types. Other ubiquitous promoters such as the CBA promoter, a fusion of the chicken-actin promoter and CMV immediate-early enhancer, allows stable GFP reporter expression in both RPE and photoreceptor cells after subretinal injections.

Modulation of Expression

Sometimes modulation of transgene expression may be necessary since strong constitutive expression of a therapeutic gene in retinal tissues could be deleterious for long-term retinal function. Different methods have been utilized for the expression modulation. One way is using exogenously regulatable promoter system in AAV vectors. For example, the tetracycline-inducible expression system uses a silencer/transactivator

AAV2 vector and a separate inducible doxycline-responsive coinjection. When induction occurs by oral doxycycline, this system shows tight regulation of gene expression in both photoreceptor and RPE cells.

Examples and Animal Models

Targeting RPE

One study that was done by Royal College of Surgeons (RCS) in rat model shows that a recessive mutation in a receptor tyrposine kinase gene, mertk results in a premature stop codon and impaired phagocytosis function by RPE cells. This mutation causes the accumulation of outer segment debris in the subretinal space, which causes photoreceptor cell death. The model organism with this disease received a subretinal injection of AAV serotype 2 carrying a mouse Mertk cDNA under the control of either the CMV or RPE65 promoters. This treatment was found to prolong photoreceptor cell survival for several months and also the number of photoreceptor was 2.5 fold higher in AAV-Mertk- treated eyes compared with controls 9 weeks after injection, also they found decreased amount of debris in the subretinal space.

The protein RPE65 is used in the retinoid cycle where the all-trans-retinol within the rod outer segment is isomerized to its 11-cis form and oxidized to 11-cis retinal before it goes back to the photoreceptor and joins with opsin molecule to form functional rhodopsin. In animal knockout model (RPE65-/-), gene transfer experiment shows that early intraocular delivery of human RPE65 vector on embryonic day 14 shows efficient transduction of retinal pigment epithelium in the RPE65-/- knockout mice and rescues visual functions. This shows successful gene therapy can be attributed to early intraocular deliver to the diseased animal.

Targeting of Photoreceptors

Juvenile retinoschisis is a disease that affects the nerve tissue in the eye. This disease is an X-linked recessive degenerative disease of the central macula region, and it is caused by mutation in the RSI gene encoding the protein retinoschisin. Retinoschisin is produced in the photoreceptor and bipolar cells and it is critical in maintaining the synaptic integrity of the retina.

Specifically the AAV 5 vector containing the wild-type human RSI cDNA driven by a mouse opsin promoter showed long-term retinal functional and structural recovery. Also the retinal structural reliability improved greatly after the treatment, characterized by an increase in the outer nuclear layer thickness.

Retinitis Pigmentosa

Retinitis pigmentosa is an inherited disease which leads to progressive night blindness and loss of peripheral vision as a result of photoreceptor cell death. Most people who suffer from RP are born with rod cells that are either dead or dysfunctional, so they are effectively blind at nighttime, since these are the cells responsible for vision in low levels of light. What follows often is the death of cone cells, responsible for color vision and acuity, at light levels present during the day. Loss of cones leads to full blindness as early as five years old, but may not onset until many years later. There have been multiple hypotheses about how the lack of rod cells can lead to the death of cone cells. Pinpointing a mechanism for RP is difficult because there are more than 39 genetic loci and genes correlated with this disease. In an effort to find the cause of RP, there have been different gene therapy techniques applied to address each of the hypotheses.

Different types of inheritance can attribute to this disease; autosomal recessive, autosomal dominant, X-linked type, etc. The main function of rhodopsin is initiating the phototransduction cascade. The opsin proteins are made in the photoreceptor inner segments, then transported to the outer segment, and eventually phagocytized by the RPE cells. When mutations occur in the rhodopsin the directional protein movement is affected because the mutations can affect protein folding, stability, and intracellular trafficking. One approach is introducing AAV-delivered ribozymes designed to target and destroy a mutant mRNA.

The way this system operates was shown in animal model that have a mutant rhodopsin gene. The injected AAV-ribozymes were optimized in vitro and used to cleave the mutant mRNA transcript of P23H (where most mutation occur) in vivo.

Another mutation in the rhodopsin structural protein, specifically peripherin 2 which is a membrane glycoprotein involved in the formation of photoreceptor outersegment disk, can lead to recessive RP and macular degeneration in human (19). In a mouse experiment, AAV2 carrying a wild-type peripherin 2 gene driven by a rhodopsin promoter was delivered to the mice by subretinal injection. The result showed improvement in photoreceptor structure and function which was detected by ERG (electroretinogram). The result showed improvement of photoreceptor structure and function which was detected by ERG. Also peripherin 2 was detected at the outer segment layer of the retina 2 weeks after injection and therapeutic effects were noted as soon as 3 weeks after injection. Interestingly, a well-defined outer segment containing both peripherin2 and rhodopsin was present 9-month after injection.

Since apoptosis can be the cause of photoreceptor death in most of the retinal dystrophies. It has been known that survival factors and antiapoptoic reagents can be an alternative treatment if the mutation is unknown for gene replacement therapy. Some scientists have experimented with treating this issue by injecting substitute trophic factors into the eye. One group of researchers injected the rod derived cone viability factor (RdCVF) protein (encoded for by the Nxnl1 (Txnl6) gene) into the eye of the most commonly occurring dominant RP mutation rat models. This treatment demonstrated success in promoting the survival of cone activity, but the treatment served even more significantly to prevent progression of the disease by increasing the actual function of the cones. Experiments were also carried out to study whether supplying AAV2 vectors with cDNA for glial cell line-derived neurotrophic factor (GDNF) can have an anti-apoptosis effect on the rod cells. In looking at an animal model, the opsin transgene contains a truncated protein lacking the last 15 amino acids of the C terminus, which causes alteration in rhodopsin transport to the outer segment and leads to retinal degeneration When the AAV2-CBA-GDNF vector is administered to the subretinal space, photoreceptor stabilized and rod photoreceptors increased and this was seen in the improved function of the ERG analysis. Successful experiments in animals have also been carried out using ciliary neurotrophic factor (CNTF), and CNTF is currently being used as a treatment in human clinical trials.

AAV-based Treatment for Retinal Neovascular Diseases

Ocular neovascularization (NV) is the abnormal formation of new capillaries from already existing blood vessels in the eye, and this is a characteristics for ocular diseases such as diabetic retinopathy (DR), retinopathy of prematurity (ROP) and (wet form) age-related macular degeneration (AMD). One of the main players in these diseases is VEGF (Vascular endothelial growth factor) which is known to induce vessel leakage and which is also known to be angiogenic. In normal tissues VEGF stimulates endothelial cell proliferation in a dose dependent manner, but such activity is lost with other angiogenic factors.

Many angiostatic factors have been shown to counteract the effect of increasing local VEGF. The naturally occurring form of soluble Flt-1 has been shown to reverse neovascularization in rats, mice, and monkeys.

Pigment epithelium-derived factor (PEDF) also acts as an inhibitor of angiogenesis. The secretion of PEDF is noticeably decreased under hypoxic conditions allowing the endothelial mitogenic activity of VEGF to dominate, suggesting that the loss of PEDF plays a central role in the development of ischemia-driven NV. One interesting clinical finding shows that the levels of PEDF in aqueous humor of human are decreased with increasing age, indicating that the reduction may lead to the development of AMD. In animal model, an AAV with human PEDF cDNA under the control of the CMV promoter prevented choroidal and retinal NV (24).

The finding suggests that the AAV-mediated expression of angiostatic factors can be implemented to treat NV. This approach could be useful as an alternative to frequent injections of recombinant protein into the eye. In addition, PEDF and sFlt-1 may be able to diffuse through sclera tissue, allowing for the potential to be relatively independent of the intraocular site of administration.

References

- Gould, D (2012). "Gene doping: Gene delivery for olympic victory". British Journal of Clinical Pharmacology. 76 (2): 292–298. PMC 3731603 . PMID 23082866. doi:10.1111/bcp.12010

- Retarded, Neil A.; Reece, Jane B.; Urry, Lisa A.; Cain, Michael L.; Wasserman, Steven A.; Minorsky, Peter V.; Jackson, Robert B. (2009). Biology (9th ed.). p. 229. ISBN 978-0-8053-6844-4
- Gallagher, James. (2 November 2012) BBC News – Gene therapy: Glybera approved by European Commission. BBC. Retrieved 15 December 2012
- Birzniece, V (2015). "Doping in sport: Effects, harm and misconceptions". Internal Medicine Journal. 45 (3): 239–48. PMID 25369881. doi:10.1111/imj.12629
- Pieter Dirk Nieuwkoop; Lien A. Sutasurya (1979). Primordial Germ Cells in the Chordates: Embryogenesis and Phylogenesis. CUP Archive. ISBN 978-0-521-22303-4
- Garde, Damian (6 November 2014) Spark snags a 'breakthrough' tag for its Phase III gene therapy FierceBiotech, Retrieved 9 February 2015
- Cieszczyk, Pawel; MacIejewska, Agnieszka; Sawczuk, Marek (2009). "Gene doping in modern sport" (PDF). Journal Biology of Exercise. 5. doi:10.4127/jbe.2009.0021
- Juan J. Tarin; Antonio Cano (14 September 2000). Fertilization in Protozoa and Metazoan Animals: Cellular and Molecular Aspects. Springer. ISBN 978-3-540-67093-3
- "Safety and Efficacy Study of rAAV.sFlt-1 in Patients With Exudative Age-Related Macular Degeneration (AMD)". U. S. National Institutes of Health. Retrieved 1 June 2012
- Andrew Lowe; Stephen Harris; Paul Ashton (1 April 2009). Ecological Genetics: Design, Analysis, and Application. John Wiley & Sons. pp. 108–. ISBN 978-1-4443-1121-1
- "Safety and Tolerability Study of AAV2-sFLT01 in Patients With Neovascular Age-Related Macular Degeneration (AMD)". U. S. National Institutes of Health. Retrieved 1 June 2012
- Lebherz C.; Maguire A.; Tang W.; Bennett J.; Wilson J. M. (2008). "Novel AAV serotypes for improved ocular gene transfer". The Journal of Gene Medicine. 10 (4): 375–382. doi:10.1002/jgm.1126
- Nikolas Zagris; Anne Marie Duprat; Antony Durston (30 November 1995). Organization of the Early Vertebrate Embryo. Springer. pp. 2–. ISBN 978-0-306-45132-4
- "First Patient Treated in Choroideremia Gene Therapy Clinical Trial in U.K.". Foundation Fighting Blindness. 28 October 2011. Retrieved 1 June 2012
- Cepko, C. L. (2012). "Emerging Gene Therapies for Retinal Degenerations". Journal of Neuroscience. 32 (19): 6415–6420. doi:10.1523/JNEUROSCI.0295-12.2012. PMC 3392151. PMID 22573664

3

In Situ and *In Vivo* Gene Therapy: An Integrated Study

In vivo gene therapy involves the introduction of biomolecules directly into the patient's blood stream so that the vector may deliver the therapeutic agent by finding and reaching the target tissue. *In situ* gene therapy involves the direct placement of corrected genes onto the affected area.This chapter discusses the methods of *in situ* and *in vivo* gene therapy in a critical manner providing key analysis to the subject matter.

In Situ

In situ is a Latin phrase that translates literally to "on site" or "in position". It means "locally", "on site", "on the premises" or "in place" to describe an event where it takes place, and is used in many different contexts. For example, in fields such as physics, chemistry, or biology, *in situ* may describe the way a measurement is taken, that is, in the same place the phenomenon is occurring without isolating it from other systems or altering the original conditions of the test.

Aerospace

In the aerospace industry, equipment on-board aircraft must be tested *in situ*, or in place, to confirm everything functions properly as a system. Individually, each piece may work but interference from nearby equipment may create unanticipated problems. Special test equipment is available for this *in situ* testing.

Archaeology

In archaeology, *in situ* refers to an artifact that has not been moved from its original place of deposition. In other words, it is stationary, meaning "still." An artifact being *in situ* is critical to the interpretation of that artifact and, consequently, of the culture which formed it. Once an artifact's 'find-site' has been recorded, the artifact can then be moved for conservation, further interpretation and display. An artifact that is not discovered *in situ* is considered out of context and as not providing an accurate picture of the associated culture. However, the out of context artifact can provide scientists with an example of types and locations of *in situ* artifacts yet to be discovered. When excavating a burial site or surface deposit "in situ" refers to cataloging, recording, mapping, photographing human remains in the position they are discovered.

Ancient Hohokam arrowhead *in situ*.

The label *in situ* indicates only that the object has not been "newly" moved. Thus, an archaeological *in situ* find may be an object that was historically looted from another place, an item of "booty" of a past war, a traded item, or otherwise of foreign origin. Consequently, the *in situ* find site may still not reveal its provenance, but with further detective work may help uncover links that otherwise would remain unknown. It is also possible for archaeological layers to be reworked on purpose or by accident (by humans, natural forces or animals). For example, in a Tell mound, where layers are not typically uniform or horizontal, or in land cleared or tilled for farming.

The term *in situ* is often used to describe ancient sculpture that was carved in place such as the Sphinx or Petra. This distinguishes it from statues that were carved and moved like the Colossi of Memnon, which was moved in ancient times.

Art

In art, *in situ* refers to a work of art made specifically for a host site, or that a work of art takes into account the site in which it is installed or exhibited. For a more detailed account see: Site-specific art. The term can also refer to a work of art created at the site where it is to be displayed, rather than one created in the artist's studio and then installed elsewhere (*e.g.*, a sculpture carved *in situ*). In architectural sculpture the term is frequently employed to describe sculpture that is carved on a building, frequently from scaffolds, after the building has been erected.

Astronomy

A fraction of the globular star clusters in our galaxy, as well as those in other massive galaxies, might have formed *in situ*. The rest might have been accreted from now defunct dwarf galaxies.

In astronomy, *in situ* also refers to *in situ* planet formation, in which planets are hypothesized to have been formed in the orbit that they are currently observed to be in rather than migrating from a different orbit.

Biology and Biomedical Engineering

In biology and biomedical engineering, *in situ* means to examine the phenomenon exactly in place where it occurs (i.e., without moving it to some special medium).

Live individual of the sea snail *Nataea* photographed *in situ*

In the case of observations or photographs of living animals, it means that the organism was observed (and photographed) in the wild, exactly as it was found and exactly where it was found. This means it was not taken out of the area. The organism had not been moved to another (perhaps more convenient) location such as an aquarium.

This phrase *in situ* when used in laboratory science such as cell science can mean something intermediate between *in vivo* and *in vitro*. For example, examining a cell within a whole organ intact and under perfusion may be *in situ* investigation. This would not be *in vivo* as the donor is sacrificed by experimentation, but it would not be the same as working with the cell alone (a common scenario for *in vitro* experiments).

In vitro was among the first attempts to qualitatively and quantitatively analyze natural occurrences in the lab. Eventually, the limitation of *in vitro* experimentation was that they were not conducted in natural environments. To compensate for this problem, *in vivo* experimentation allowed testing to occur in the original organism or environment. To bridge the dichotomy of benefits associated with both methodologies, *in situ* experimentation allowed the controlled aspects of *in vitro* to become coalesced with the natural environmental compositions of *in vivo* experimentation.

In conservation of genetic resources, "*in situ* conservation" (also "on-site conservation") is the process of protecting an endangered plant or animal species in its natural habitat, as opposed to *ex situ* conservation (also "off-site conservation").

Chemistry and Chemical Engineering

In chemistry, *in situ* typically means "in the reaction mixture."

There are numerous situations in which chemical intermediates are synthesized *in situ* in various processes. This may be done because the species is unstable, and cannot be isolated, or simply out of convenience. Examples of the former include the Corey-Chaykovsky reagent and adrenochrome.

In biomedical engineering, protein nanogels made by the in situ polymerization method provide a versatile platform for storage and release of therapeutic proteins. It has tremendous applications for cancer treatment, vaccination, diagnosis, regenerative medicine, and therapies for loss-of-function genetic diseases.

In chemical engineering, *in situ* often refers to industrial plant "operations or procedures that are performed in place". For example, aged catalysts in industrial reactors may be regenerated in place (*in situ*) without being removed from the reactors.

Civil Engineering

In architecture and building, *in situ* refers to construction which is carried out at the building site using raw materials. Compare that with *prefabricated* construction, in which building components are made in a factory and then transported to the building site for assembly. For example, concrete slabs may be *in situ* (also "cast-in-place") or *prefabricated.*

In situ techniques are often more labour-intensive, and take longer, but the materials are cheaper, and the work is versatile and adaptable. *Prefabricated* techniques are usually much quicker, therefore saving money on labour costs, but factory-made parts can be expensive. They are also inflexible, and must often be designed on a grid, with all details fully calculated in advance. Finished units may require special handling due to excessive dimensions.

The phrase may also refer to those assets which are present at or near a project site. In this case, it is used to designate the state of an unmodified sample taken from a given stockpile.

Site construction usually involves grading the existing soil surface so that material is "cut" out of one area and "filled" in another area creating a flat pad on an existing slope. The term "in situ" distinguishes soil still in its existing condition from soil modified (filled) during construction. The differences in the soil properties for supporting foundations, accepting underground utilities, and shedding storm water persist indefinitely.

Computer Science

In computer science an *in situ* operation is one that occurs without interrupting the normal state of a system. For example, a file backup may be restored over a running system, without needing to take the system down to perform the restore. In the context of a database, a restore would allow the database system to continue to be available to users while a restore happened. An *in situ* upgrade would allow an operating system, firmware or application to be upgraded while the system was still running, perhaps without the need to reboot it, depending on the sophistication of the system.

An algorithm is said to be an *in situ* algorithm, or in-place algorithm, if the extra amount of memory required to execute the algorithm is O(1), that is, does not exceed a constant no matter how large the input. For example, heapsort is an *in situ* sorting algorithm.

In designing user interfaces, the term *in situ* means that a particular user action can be performed without going to another window, for example, if a word processor displays an image and allows you to edit the image without launching a separate image editor, this is called *in situ editing*.

In Big Data space, in situ data would mean bringing the computation to where data is located, rather than the other way. like in traditional RDBMS systems where data is moved to computational space.

Design and Advertising

In design and advertising the term typically means the superimposing of theoretical design elements onto photographs of real world locations. This is a pre-visualization tool to aid in illustrating a proof of concept.

Earth and Atmospheric Sciences

In physical geography and the Earth sciences, *in situ* typically describes natural material or processes prior to transport. For example, *in situ* is used in relation to the distinction between weathering and erosion, the difference being that erosion requires a transport medium (such as wind, ice, or water), whereas weathering occurs *in situ*. Geochemical processes are also often described as occurring to material *in situ*.

In the atmospheric sciences, *in situ* refers to obtained through direct contact with the respective subject, such as a radiosonde measuring a parcel of air or an anemometer measuring wind, as opposed to remote sensing such as weather radar or satellites.

Economics

In economics, *in situ* is used when referring to the *in place* storage of a product, usually a natural resource. More generally, it refers to any situation where there is no out-of-pocket cost to store the product so that the only storage cost is the opportunity cost of waiting longer to get your money when the product is eventually sold. Examples of *in situ* storage would be oil and gas wells, all types of mineral and gem mines, stone quarries, timber that has reached an age where it could be harvested, and agricultural products that do not need a physical storage facility such as hay.

Electrochemistry

In electrochemistry, the phrase in situ refers to performing electrochemical experiments under operating conditions of the electrochemical cell, i.e., under potential control. This is opposed to doing ex situ experiments that are performed under the absence of potential control. Potential control preserves the electrochemical environment essential to maintain the double layer structure intact and the electron transfer reactions occurring at that particular potential in the electrode/electrolyte interphasial region.

Environmental Remediation

In situ can refer to where a clean up or remediation of a polluted site is performed using and simulating the natural processes in the soil, contrary to *ex situ* where contaminated soil is excavated and cleaned elsewhere, off site.

Experimental Physics

In experimental physics *in situ* typically refers to a method of data collection or manipulation of a sample without exposure to an external environment. For example, the Si(111) 7x7 surface reconstruction is visible using a scanning tunneling microscope when it is prepared and analyzed *in situ*.

Experimental Psychology

In psychology experiments, *in situ* typically refers to those experiments done in a field setting as opposed to a laboratory setting.

Gastronomy

In gastronomy, "in situ" refers to the art of cooking with the different resources that are available on the site of the event. Here you are not going to the restaurant, but the restaurant comes to your home.

Law

In legal contexts, *in situ* is often used for its literal meaning. For example, in Hong Kong, "*in situ* land exchange" involves the government exchanging the original or expired lease of a piece of land with a new grant or re-grant with the same piece of land or a portion of that.

In the field of recognition of governments under public international law the term *in situ* is used to distinguish between an exiled government and a government with effective control over the territory, i.e. the government *in situ*.

Linguistics

In linguistics, specifically syntax, an element may be said to be *in situ* if it is pronounced in the position where it is interpreted. For example, questions in languages such as Chinese have *in situ* wh-elements, with structures comparable to "John bought what?" with *what* in the same position in the sentence as the grammatical object would be in its affirmative counterpart (for example, "John bought bread"). An example of an English wh-element that is not *in situ*: "What did John buy?"

Literature

In literature *in situ* is used to describe a condition. The Rosetta Stone, for example, was originally erected in a courtyard, for public viewing. Most pictures of the famous stone are not *in situ* pictures of it erected, as it would have been originally. The stone was uncovered as part of building material, within a wall. Its in situ condition today is that it is erected, vertically, on public display at the British Museum in London, England.

Medicine

In oncology: for a carcinoma, *in situ* means that malignant cells are present as a tumor but have not metastasized, or invaded, beyond the basement membrane of where the tumor was discovered. This can happen anywhere in the body, such as the skin, breast tissue, or lung. This type of tumor can often, depending on where it is located, be removed by surgery.

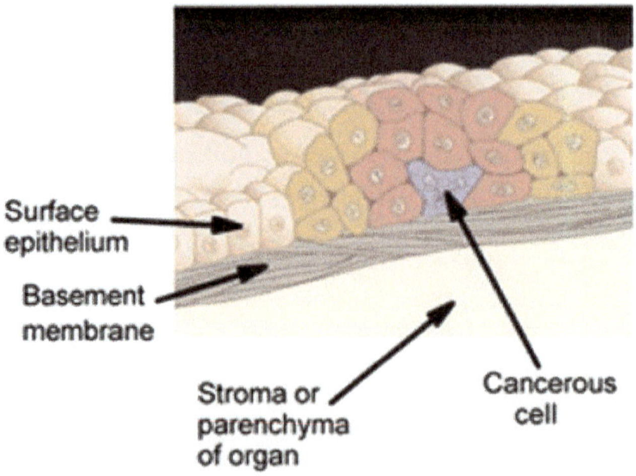

In addition to benign tumors, there are in situ tumors and invasive tumors. In situ tumors do not invade the basement membrane, whereas invasive tumors do invade the basement membrane.
Diagram of an in situ carcinoma

In anatomy: *in situ* refers to viewing structures as they appear in normal healthy bodies. For example, one can open up a cadaver's abdominal cavity and view the liver *in situ* or one can look at an isolated liver that has been removed from the cadaver's body.

In nursing, "in situ" describes any devices or appliances on the patient's body that remain in their desired and optimal position.

In medical simulation, "in situ" refers to the practice of clinical professionals using high fidelity patient simulators to train for clinical practice in patient care environments, such as wards, operating rooms, and other settings, rather than in dedicated simulation training facilities.

In biomedical, protein nanogels made by the in situ polymerization method provide a versatile platform for storage and release of therapeutic proteins. It has tremendous applications for cancer treatment, vaccination, diagnosis, regenerative medicine, and therapies for loss-of-function genetic diseases.

Mining

In situ leaching or *in situ recovery* refers to the mining technique of injecting water underground to dissolve ore and bringing the uranium-impregnated water to the surface for extraction.

Petroleum Production

In situ refers to recovery techniques which apply heat or solvents to heavy crude oil or bitumen reservoirs beneath the earth's crust. There are several varieties of *in situ* techniques, but the ones which work best in the oil sands use heat (steam).

The most common type of *in situ* petroleum production is referred to as SAGD (steam-assisted gravity drainage) this is becoming very popular in the Alberta Oil Sands.

RF Transmission

In radio frequency (RF) transmission systems, *in situ* is often used to describe the location of various components while the system is in its standard transmission mode, rather than operation in a test mode. For example, if an *in situ* wattmeter is used in a commercial broadcast transmission system, the wattmeter can accurately measure power while the station is "on air".

Space-related

Future space exploration or terraforming may rely on obtaining supplies *in situ*, such as previous plans to power the Orion space vehicle with fuel minable on the moon. The Mars Direct mission concept is based primarily on the *in situ* fuel production using Sabatier reaction.

In the space sciences, *in situ* refers to measurements of the particle and field environment that the satellite is embedded in, such as the detection of energetic particles in the solar wind, or magnetic field measurements from a magnetometer.

Urban Planning

In urban planning, in-situ upgrading is an approach to and method of upgrading informal settlements.

Vacuum Technology

In vacuum technology, *in situ* baking refers to heating parts of the vacuum system while they are under vacuum in order to drive off volatile substances that may be absorbed or adsorbed on the walls so they cannot cause outgassing.

Road Assistance

The term *in situ*, used as "repair in situ", means to repair a vehicle at the place where it has a breakdown.

In Situ Conservation

In-situ conservation is the on-site conservation or the conservation of genetic resources in natural populations of plant or animal species, such as forest genetic resources in natural populations of tree species. It is the process of protecting an endangered plant or animal species in its natural habitat, either by protecting or cleaning up the habitat itself, or by defending the species from predators. It is applied to conservation of agricultural biodiversity in agroecosystems by farmers, especially those using unconventional farming practices. e.g., Nilgiri biosphere in India.

Methods

Around 4% of the total geographical area of the country is used for *in situ* conservation. The following methods are presently used for *in situ* conservation.

In-situ conservation	Number available
Biosphere reserves	18
National parks	104
Wild-life sanctuaries	537

Biosphere Reserves

Biosphere reserves cover very large areas, often more than 5000 km^2. They are used to protect species for a long time.

Name	State
Nanda Devi	Uttrakhand
Nokrek	Meghalaya
Manas	Assam

National Parks

A national park is an area dedicated for the conservation of wildlife along with its environment. It is usually a small reserve covering an area of about 100 to 500 square kilometers. Within biosphere reserves, one or more national parks may also exist.

Name	State	Important wildlife
Kaziranga	Assam	One-horned rhino
Gir National Park	Gujarat	Asiatic lions
Bandipur	Karnataka	Tiger, Elephant
Dachigam	J & K	Hangul
Kanha	M.P	Tiger
Periyar	Kerala	Tiger, elephant

Wild Sanctuaries

A wild sanctuary is an area which is reserved for the conservation of animals only. Currently, there are 492 wild sanctuaries.

Name	State	Major wildlife
Hazaribagh sanctuary	Jharkhand	Tiger, leopard
Ghana Bird sanctuary	Rajasthan	300 species of Birds
Sultanpur Bird Sanctuary	Haryana	Migratory birds
Abohar Wild life Sanctuary	Punjab	Black buck
Nal sarovar Bird Sanctuary	Gujarat	Water birds
Mudumalai Wild life Sanctuary	Tamil Nadu	Tiger, elephant, leopard
Vedanthangal Bird Sanctuary	Tamil Nadu	Water birds

Gene Sanctuary

A gene sanctuary is an area where plants are conserved. It includes both biosphere reserve as well as national park. India has set up its first gene sanctuary in the Garo Hills of Meghalaya for wild relatives of citrus. Efforts are also being made to set up gene sanctuaries for banana, sugarcane, rice and mango.

Community Reserves

It is the type of protected area introduced in Wildlife Protection Amendment Act 2002 to provide legal support to community or privately owned reserves which cannot be designated as national park or wildlife sanctuary.

Sacred Groves

They are tract of forest set aside where all the trees and wildlife within are venerated and given total protection.

Benefits

One benefit of *in situ* conservation is that it maintains recovering populations in the environment where they have developed their distinctive properties. Another benefit is that this strategy helps ensure the ongoing processes of evolution and adaptation within their environments. As a last resort, ex-situ conservation may be used on some or all of the population, when *in situ* conservation is too difficult, or impossible. The species gets adjusted to the natural disasters like drought, floods, forest fires and this method is very cheap and convenient.

Reserves

Wildlife and livestock conservation is mostly based on *in situ* conservation. This involves the protection of wildlife habitats. Also, sufficiently large reserves are maintained to enable the target species to exist in large numbers. The population size must be sufficient to enable the necessary genetic diversity to survive within the population, so that it has a good chance of continuing to adapt and evolve over time. This reserve size can be calculated for target species by examining the population density in naturally occurring situations. The reserves must then be protected from intrusion or destruction by man, and against other catastrophes.

Agriculture

In agriculture, *in situ conservation* techniques are an effective way to improve, maintain, and use traditional or native varieties of agricultural crops. Such methodologies link the positive output of scientific research with farmers' experience and field work.

First, the accessions of a variety stored at a germplasm bank and those of the same variety multiplied by farmers are jointly tested in the producers field and in the laboratory, under different situations and stresses. Thus, the scientific knowledge about the production characteristics of the native varieties is enhanced. Later, the best tested accessions are crossed, mixed, and multiplied under replicable situations. At last, these improved accessions are supplied to the producers. Thus, farmers are enabled to crop improved selections of their own varieties, instead of being lured to substitute their own varieties with commercial ones or to abandon their crop. This technique of conservation of agricultural biodiversity is more successful in marginal areas, where commercial varieties are not expedient, due to climate and soil fertility constraints. Or where the taste and cooking characteristics of traditional varieties compensate for their lower yields.

In Situ and In Vivo Gene Therapy

Somatic cell gene therapy involves the transfer of gene to a diseased somatic cell either within the body or outside the body with the help of a viral or non viral gene therapy vector.

Ex vivo is any procedure accomplished outside. In gene therapy clinical trials cells are modified in a variety of ways to correct the gene. In *ex vivo* cells are modified outside the patient's body and the corrected version is transplanted back in to the patient. The cells are treated with either a viral or non viral gene therapy vector carrying the corrected copy of the gene.

Opposite of *ex vivo* is what we call *in vivo* where cells are treated inside the patient's body. The corrected copy of the genes is transferred into the body of the patient. The cells may be treated either with a viral or non viral vector carrying the corrected copy of the gene. If the patient is weak or the cell cannot be extracted out from the body, the gene is introduced directly into the body.

Gene therapy done in a restricted area or to a particular site is called *in-situ* . *In situ* gene therapy requires the vector to be placed directly into the affected tissues. *In vivo* gene therapy involves injecting the vector into the blood stream. The vector then must find the target tissue and deliver the therapeutic genes.

In Situ Gene therapy

In situ gene therapy comprises transfer of corrected copy of the gene into the targeted organ or tissue. The major concern of current time gene therapy protocol is the lack of efficient transduction of the targeted organ. The method is effectively used against cystic fibrosis, a disease of airway epithelium (discussed in later chapters). The method is also explored for cancer gene therapy where the viral vector is engineered to contain the herpes simplex virus thymidine kinase gene. After injection of the viral vector the patient is treated with a prodrug such as Ganciclovir, which causes 75% reduction in the tumor cell population.

In Vivo Gene Therapy

Delivery of corrected copy of the gene systemically through injection is a highly efficient way to transfer a transgene to the patient's body. The major problem of *in vivo* method is its inefficient targeting. The transgene delivered into the body by means of viral or non viral vector also evokes the immune response. The immune response against the vector leads to its clearance and only transient expression of transgene. The neutralizing antibody does not allow the second injection of the vector. Reducing the neutralizing antibody is the current area of research in order to improve the delivery of gene therapy vector. All gene therapy delivery protocols require the transgene to cross the plasma membrane and enter inside the nucleus. The major obstacle is still to deliver the transgene effectively to the intracellular compartment. Many modifications have been suggested into the viral vectors and also non viral vectors to target the gene to the tissue. VP22, a protein of herpes simplex virus has a property to spread from one cell to the other, and this property has been successfully implemented in designing the vectors.

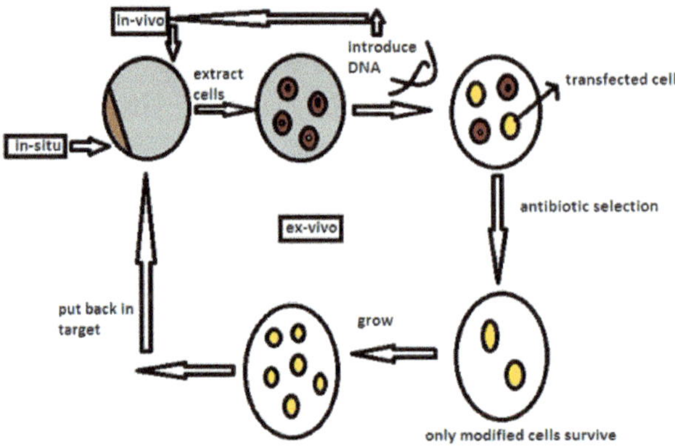

Ex vivo, in vivo, and *in situ* gene therapy:

In vivo gene therapy can be used as direct injection or by the use of biolistic gene guns. The transgene can also be transferred by means of viral or non-viral vectors. The recombinant viruses are designed in such a way as to carry the transgene with tissue specific promoter. Alternatively, DNA containing the transgene may be injected along with liposome in the patients. The plasmid DNA is also targeted with the help of mechanical means such as gene gun.

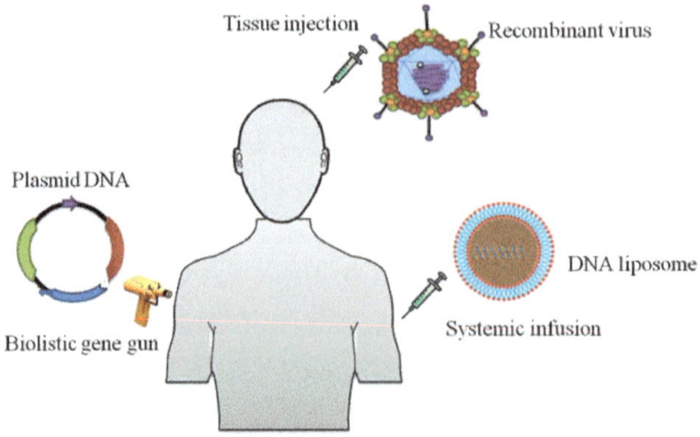

In vivo gene therapy approaches:

In vivo, the transgene can be directly injected by various routes based on the clinical condition of the patient and design of the vector. For instance, transducing the lung airway epithelium is targeted. The target specific gene therapy is possible with the help of unique promoters. The liver directed gene therapy is done usually by thyroid binding globulin promoter (TBG) while cytomegalovirus (CMV) promoter can be used to direct the muscle cells. *Ex vivo* on the other hand requires the manipulation of the cells outside the body and the transducer cells are injected back after quality assurance.

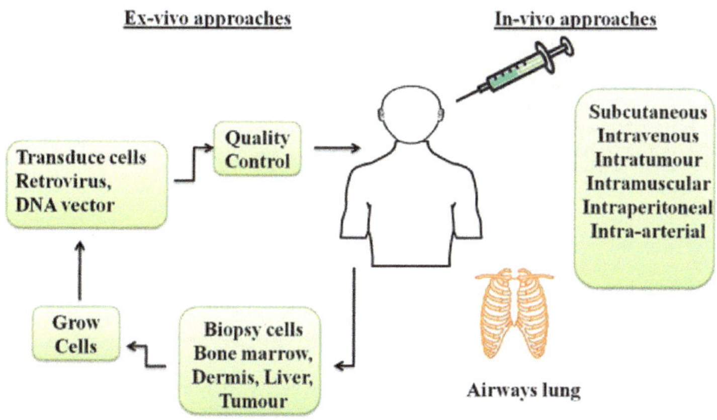

Clinical *ex vivo* and *in vivo* approach

In Vivo

Studies that are *in vivo* (Latin for "within the living"; often not italicized in English) are those in which the effects of various biological entities are tested on whole, living organisms, usually animals, including humans, and plants as opposed to a partial or dead organism. This is not to be confused with experiments done *in vitro* ("within the glass"), i.e., in a laboratory environment using test tubes, petri dishes, etc. Examples of investigations *in vivo* include: the pathogenesis of disease by comparing the effects of bacterial infection with the effects of purified bacterial toxins; the development of antibiotics, antiviral drugs, and new drugs generally; and new surgical procedures. Consequently, animal testing and clinical trials are major elements of *in vivo* research. *In vivo* testing is often employed over *in vitro* because it is better suited for observing the overall effects of an experiment on a living subject. In drug discovery, for example, verification of efficacy *in vivo* is crucial, because *in vitro* assays can sometimes yield misleading results with drug candidate molecules that are irrelevant *in vivo* (e.g., because such molecules cannot reach their site of *in vivo* action, for example as a result of rapid catabolism in the liver).

The English microbiologist Professor Harry Smith and his colleagues in the mid-1950s showed the importance of *in vivo* studies. They found that sterile filtrates of serum from animals infected with *Bacillus anthracis* were lethal for other animals, whereas extracts of culture fluid from the same organism grown *in vitro* were not. This discovery of anthrax toxin through the use of *in vivo* experiments had a major impact on studies of the pathogenesis of infectious disease.

The maxim *in vivo veritas* ("in a living thing [there is] truth") is used to describe this type of testing and is a play on *in vino veritas*, ("in wine [there is] truth"), a well-known proverb.

In Vivo Vs. Ex Vivo Research

In microbiology, *in vivo* is often used to refer to experimentation done in live isolated cells rather than in a whole organism, for example, cultured cells derived from biopsies. In this situation, the more specific term is *ex vivo*. Once cells are disrupted and individual parts are tested or analyzed, this is known as *in vitro*.

Methods of use

According to Christopher Lipinski and Andrew Hopkins, "Whether the aim is to discover drugs or to gain knowledge of biological systems, the nature and properties of a chemical tool cannot be considered independently of the system it is to be tested in. Compounds that bind to isolated recombinant proteins are one thing; chemical tools that can perturb cell function another; and pharmacological agents that can be tolerated by a live organism and perturb its systems are yet another. If it were simple to ascertain the properties required to develop a lead discovered *in vitro* to one that is active *in vivo*, drug discovery would be as reliable as drug manufacturing."

Ex Vivo

Ex vivo (Latin: "out of the living") means that which takes place outside an organism. In science, *ex vivo* refers to experimentation or measurements done in or on tissue from an organism in an external environment with minimal alteration of natural conditions. *Ex vivo* conditions allow experimentation on an organism's cells or tissues under more controlled conditions than is possible in *in vivo* experiments (in the intact organism), at the expense of altering the "natural" environment.

A primary advantage of using *ex vivo* tissues is the ability to perform tests or measurements that would otherwise not be possible or ethical in living subjects. Tissues may be removed in many ways, including in part, as whole organs, or as larger organ systems.

Examples of *ex vivo* specimen use include:

- assays;
- finding cancer treatment agents that are effective against the organism's cancer cells

- measurements of physical, thermal, electrical, mechanical, optical and other tissue properties, especially in various environments that may not be life-sustaining (for example, at extreme pressures or temperatures);
- realistic models for surgical procedure development;
- investigations into the interaction of different energy types with tissues;
- or as phantoms in imaging technique development.

The term *ex vivo* means that the samples to be tested have been extracted from the organism. The term *in vitro* ("within the glass") means the samples to be tested are obtained from a repository, in the case of cancer cells a strain that would produce favorable results, then grown to produce a control sample and the number of samples required for the number of tests. These two terms are not synonymous even though the testing in both cases is "within the glass". The results of *ex vivo* only apply to the organism supplying the cell whereas the *in vitro* results only apply to the cell line used.

In cell biology, *ex vivo* procedures often involve living cells or tissues taken from an organism and cultured in a laboratory apparatus, usually under sterile conditions with no alterations, for up to 24 hours to obtain sufficient cells for the experiments. Experiments generally start after 24 hours of incubation. Using living cells or tissue from the same organism are still considered to be *ex vivo*. One widely performed *ex vivo* study is the chick chorioallantoic membrane (CAM) assay. In this assay, angiogenesis is promoted on the CAM membrane of a chicken embryo outside the organism (chicken).

Modification by in Vivo and Ex Vivo Gene Therapy

Somatic cells are modified in a number of different ways

Gene Supplementation

This method is also called as gene augmentation. It aims to supply a functional copy of the defective gene. The method is generally employed for a gene product that has lost its function or is showing inadequate expression of protein. The process can be used when there is irreversible damage of the gene. The gene supplementation can be used for cancer therapy to increase the immune response against the tumor cells. Alternatively it can be used to replace the defective tumor suppressor gene.

Gene Replacement

In this the mutated or nonfunctional copy of the gene is replaced by the correct functional copy of the same gene. The gene replacement is performed for a mutated gene which is harmful for the host. In general gene replacement aims for gain of function.

Targeted Inhibition of Gene Function

The targeted inhibition of gene function is relevant for the infectious diseases where specific gene of pathogens is targeted. The pathogen associated antigenic gene is knocked down in order to avoid the harmful effect of the protein. It also aims for targeted inhibition of tumor antigen to reduce the autoimmune response. The gene is silenced by various means including siRNA, RNAi, etc.

Targeted Killing of the Cells

The targeted killing aims specifically for cancer cells where the metastatic form of the tumors are targeted and killed *in situ* . Many novel viruses called oncolytic viruses are targeted to kill the cancerous cells. *Paramyxoviruses* belong to such group of promising oncolytic viruses. Many studies using *paramyxoviruses* have shown encouraging results in reducing the cancerous condition by specifically targeting and killing the cancer cells by apoptosis.

Mechanism of Suicidal Gene Therapy

Retroviruses or lentiviral vectors are designed to carry a therapeutic gene. Since the retroviruses only grow in dividing cells so they will specifically multiply in tumor cells. Retroviral vectors are designed to express the thymidine kinase gene (tk) from herpes virus.

The tk gene sensitizes the tumor cells for a prodrug ganciclovir. Herepes virus tk is a normal substrate for the ganciclovir while the host tk is not affected by the drug. Therefore host cells surrounding the tumors will not have any effect of ganciclovir.

Ganciclovir is phosphorylated by the herpes virus tk into monophosphate form; the ganciclovir is then converted into triphosphate form which inhibits the DNA synthesis in tumor cells. The tumors cells are specifically killed by this process as the host tk has little affinity with the ganciclovir.

Herpes simplex virus contains an enzyme called THYMIDINE KINASE which is used in targeting cancer cells for suicidal gene therapy. The Ganciclovir, an antiviral drug is used for suicidal gene therapy. The underlying mechanism behind the killing of a cancer cells is by phosphorylation of Ganciclovir.

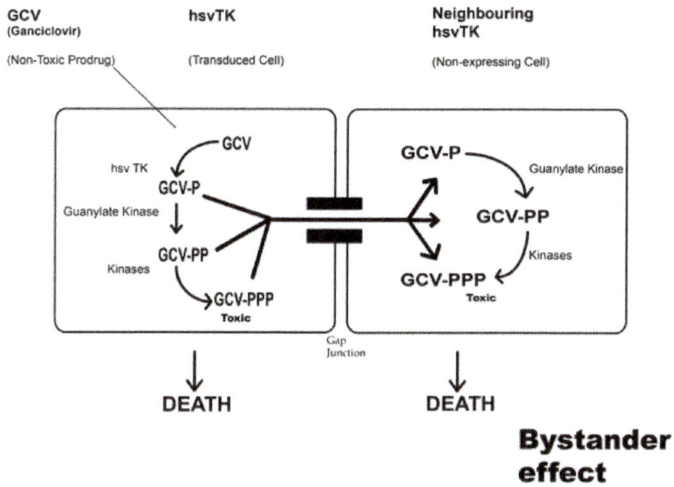

Schematic representation of suicidal gene therapy

Viral vectors are categorized into integrating or non-integrating vectors based on their recombination capacity with the host cell chromosome. Adeno-associated viruses are known to target the genetic material to human chromosome number 19 (19q13.4). The incorporation of genes into the chromosome can lead to STABLE expression of the protein of interest. The other type is called as TEMPORARY where the proteins are expressed only for a short period of time and gene is not integrated with the chromosome.

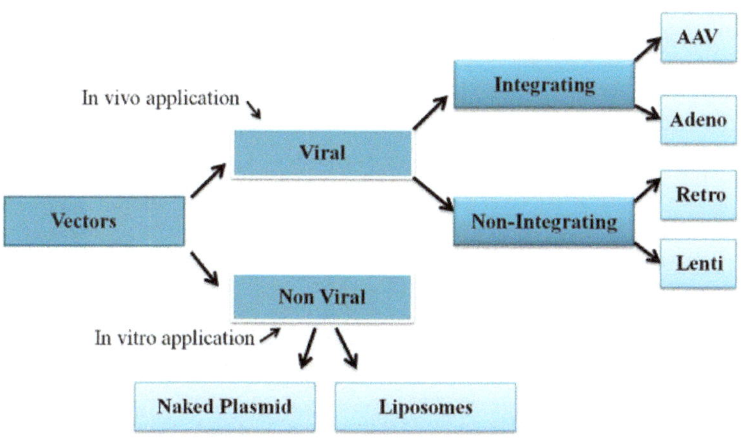

Application of viral and non viral vectors for *in vivo* and *in vitro* approach

References

- Iverson, Cheryl, et al. (eds) (2007). "12.1.1 Use of Italics". AMA Manual of Style (10th ed.). Oxford, Oxfordshire: Oxford University Press. ISBN 978-0-19-517633-9
- Chiang, E.; Laughlin, G. (2013-06-01). "The Minimum-Mass Extrasolar Nebula: In-Situ Formation of Close-In Super-Earths". Monthly Notices of the Royal Astronomical Society. 431 (4): 3444–3455. ISSN 0035-8711. arXiv:1211.1673 . doi:10.1093/mnras/stt424

- merican Psychological Association (2010), "4.21 Use of Italics", The Publication Manual of the American Psychological Association (6th ed.), Washington, DC, USA: APA, ISBN 978-1-4338-0562-2
- Ye, Yanqi; Yu, Jicheng; Gu, Zhen (2015). "Versatile Protein Nanogels Prepared by In Situ Polymerization". Macromolecular Chemistry and Physics. doi:10.1002/macp.201500296
- Gillespie, Cailein; Cousins, John A. (2001). European Gastronomy into the 21st Century. Oxford, UK: Elsevier. p. 72. ISBN 978-0-7506-5267-4. Retrieved 16 June 2014
- Lipinski C, Hopkins A (2004). "Navigating chemical space for biology and medicine". Nature. 432 (7019): 855–61. PMID 15602551. doi:10.1038/nature03193
- Elise A. Friedland; Melanie Grunow Sobocinski (3 February 2015). The Oxford Handbook of Roman Sculpture. Oxford University Press. pp. 113–. ISBN 978-0-19-026687-5
- Pamela A. Webb (1996). Hellenistic Architectural Sculpture: Figural Motifs in Western Anatolia and the Aegean Islands. Univ of Wisconsin Press. pp. 65–. ISBN 978-0-299-14980-2

4

A Comprehensive Study of Viral Vectors

The tools used by molecular biologists to introduce genetic material in cells are called viral vectors. Normally a viral vector is created according to particular needs; however, they share a few characteristics. These features include lox toxicity, stability, safe nature, identification and cell type specificity. Lentiviruses are especially used in gene therapy. This chapter has been carefully written to provide an easy understanding of the varied facets of viral vectors.

Viral Vector

Viral vectors are tools commonly used by molecular biologists to deliver genetic material into cells. This process can be performed inside a living organism (*in vivo*) or in cell culture (*in vitro*). Viruses have evolved specialized molecular mechanisms to efficiently transport their genomes inside the cells they infect. Delivery of genes by a virus is termed transduction and the infected cells are described as transduced. Molecular biologists first harnessed this machinery in the 1970s. Paul Berg used a modified SV40 virus containing DNA from the bacteriophage λ to infect monkey kidney cells maintained in culture.

Key Properties of a Viral Vector

Viral vectors are tailored to their specific applications but generally share a few key properties.

- *Safety*: Although viral vectors are occasionally created from pathogenic viruses, they are modified in such a way as to minimize the risk of handling them. This usually involves the deletion of a part of the viral genome critical for viral replication. Such a virus can efficiently infect cells but, once the infection has taken place, requires a helper virus to provide the missing proteins for production of new virions.
- *Low toxicity*: The viral vector should have a minimal effect on the physiology of the cell it infects.
- *Stability*: Some viruses are genetically unstable and can rapidly rearrange their genomes. This is detrimental to predictability and reproducibility of the work conducted using a viral vector and is avoided in their design.

- *Cell type specificity*: Most viral vectors are engineered to infect as wide a range of cell types as possible. However, sometimes the opposite is preferred. The viral receptor can be modified to target the virus to a specific kind of cell. Viruses modified in this manner are said to be pseudotyped.
- *Identification*: Viral vectors are often given certain genes that help identify which cells took up the viral genes. These genes are called Markers. A common marker is antibiotic resistance to a certain antibiotic. The cells can then be isolated easily as those that have not taken up the viral vector genes do not have antibiotic resistance and so cannot grow in a culture with antibiotics present.

Applications

Basic Research

Viral vectors were originally developed as an alternative to transfection of naked DNA for molecular genetics experiments. Compared to traditional methods such as calcium phosphate precipitation, transduction can ensure that nearly 100% of cells are infected without severely affecting cell viability. Furthermore, some viruses integrate into the cell genome facilitating stable expression.

Protein coding genes can be expressed using viral vectors, commonly to study the function of the particular protein. Viral vectors, especially retroviruses, stably expressing marker genes such as GFP are widely used to permanently label cells to track them and their progeny, for example in xenotransplantation experiments, when cells infected *in vitro* are implanted into a host animal.

Gene insertion is cheaper to carry out than gene knockout. But as the silencing is sometimes non-specific and has off-target effects on other genes, it provides less reliable results. Animal host vectors also play an important role.

Gene Therapy

Gene therapy is a technique for correcting defective genes responsible for disease development. In the future, gene therapy may provide a way to cure genetic disorders, such as severe combined immunodeficiency, cystic fibrosis or even haemophilia A. Because these diseases result from mutations in the DNA sequence for specific genes, gene therapy trials have used viruses to deliver unmutated copies of these genes to the cells of the patient's body. There have been a huge number of laboratory successes with gene therapy. However, several problems of viral gene therapy must be overcome before it gains widespread use. Immune response to viruses not only impedes the delivery of genes to target cells but can cause severe complications for the patient. In one of the early gene therapy trials in 1999 this led to the death of Jesse Gelsinger, who was treated using an adenoviral vector.

Some viral vectors, for instance gamma-retroviruses, insert their genomes at a seemingly random location on one of the host chromosomes, which can disturb the function of cellular genes and lead to cancer. In a severe combined immunodeficiency retroviral gene therapy trial conducted in 2002, four of the patients developed leukemia as a consequence of the treatment; three of the patients recovered after chemotherapy. Adeno-associated virus-based vectors are much safer in this respect as they always integrate at the same site in the human genome.

Vaccines

Viruses expressing pathogen proteins are currently being developed as vaccines against these pathogens, based on the same rationale as DNA vaccines. T-lymphocytes recognize cells infected with intracellular parasites based on the foreign proteins produced within the cell. T cell immunity is crucial for protection against viral infections and such diseases as malaria. A viral vaccine induces expression of pathogen proteins within host cells similarly to the Sabin Polio vaccine and other attenuated vaccines. However, since viral vaccines contain only a small fraction of pathogen genes, they are much safer and sporadic infection by the pathogen is impossible. Adenoviruses are being actively developed as vaccines.

Types of Viral Vectors

Retroviruses

Retroviruses are one of the mainstays of current gene therapy approaches. The recombinant retroviruses such as the Moloney murine leukemia virus have the ability to integrate into the host genome in a stable fashion. They contain a reverse transcriptase that allows integration into the host genome. They have been used in a number of FDA-approved clinical trials such as the SCID-X1 trial.

Retroviral vectors can either be replication-competent or replication-defective. Replication-defective vectors are the most common choice in studies because the viruses have had the coding regions for the genes necessary for additional rounds of virion replication and packaging replaced with other genes, or deleted. These virus are capable of infecting their target cells and delivering their viral payload, but then fail to continue the typical lytic pathway that leads to cell lysis and death.

Conversely, replication-competent viral vectors contain all necessary genes for virion synthesis, and continue to propagate themselves once infection occurs. Because the viral genome for these vectors is much lengthier, the length of the actual inserted gene of interest is limited compared to the possible length of the insert for replication-defective vectors. Depending on the viral vector, the typical maximum length of an allowable DNA insert in a replication-defective viral vector is usually about 8–10 kB. While this limits the introduction of many genomic sequences, most cDNA sequences can still be accommodated.

The primary drawback to use of retroviruses such as the Moloney retrovirus involves the requirement for cells to be actively dividing for transduction. As a result, cells such as neurons are very resistant to infection and transduction by retroviruses.

There is concern that insertional mutagenesis due to integration into the host genome might lead to cancer or leukemia. This concern remained theoretical until gene therapy for ten SCID-X1 patients using Maloney murine leukemia virus resulted in two cases of leukemia caused by activation of the LMO2 oncogene due to nearby integration of the vector.

Lentiviruses

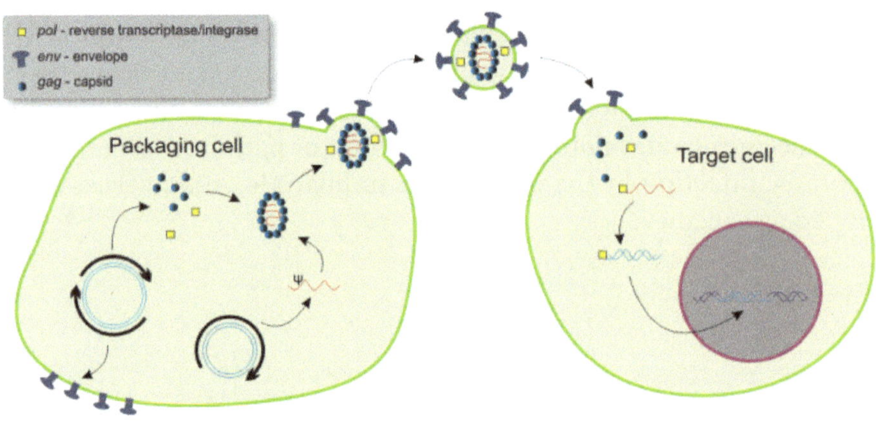

Packaging and transduction by a lentiviral vector.

Lentiviruses are a subclass of Retroviruses. They are sometimes used as vectors for gene therapy thanks to their ability to integrate into the genome of non-dividing cells, which is the unique feature of Lentiviruses as other Retroviruses can infect only dividing cells. The viral genome in the form of RNA is reverse-transcribed when the virus enters the cell to produce DNA, which is then inserted into the genome at a random position (recent findings actually suggest that the insertion of viral DNA is not random but directed to specific active genes and related to genome organisation) by the viral integrase enzyme. The vector, now called a provirus, remains in the genome and is passed on to the progeny of the cell when it divides. The site of integration is unpredictable, which can pose a problem. The provirus can disturb the function of cellular genes and lead to activation of oncogenes promoting the development of cancer, which raises concerns for possible applications of lentiviruses in gene therapy. However, studies have shown that lentivirus vectors have a lower tendency to integrate in places that potentially cause cancer than gamma-retroviral vectors. More specifically, one study found that lentiviral vectors did not cause either an increase in tumor incidence or an earlier onset of tumors in a mouse strain with a much higher incidence of tumors. Moreover, clinical trials that utilized lentiviral vectors to deliver gene therapy for the treatment of HIV experienced no increase in mutagenic or oncologic events.

For safety reasons lentiviral vectors never carry the genes required for their replication. To produce a lentivirus, several plasmids are transfected into a so-called packaging cell line, commonly HEK 293. One or more plasmids, generally referred to as packaging plasmids, encode the virion proteins, such as the capsid and the reverse transcriptase. Another plasmid contains the genetic material to be delivered by the vector. It is transcribed to produce the single-stranded RNA viral genome and is marked by the presence of the ψ (psi) sequence. This sequence is used to package the genome into the virion.

Adenoviruses

As opposed to lentiviruses, adenoviral DNA does not integrate into the genome and is not replicated during cell division. This limits their use in basic research, although adenoviral vectors are still used in *in vitro* and also *in vivo* experiments. Their primary applications are in gene therapy and vaccination. Since humans commonly come in contact with adenoviruses, which cause respiratory, gastrointestinal and eye infections, majority of patients have already developed neutralizing antibodies which can inactivate the virus before it can reach the target cell. To overcome this problem scientists are currently investigating adenoviruses that infect different species to which humans do not have immunity.

Adeno-associated Viruses

Adeno-associated virus (AAV) is a small virus that infects humans and some other primate species. AAV is not currently known to cause disease and consequently the virus causes a very mild immune response. AAV can infect both dividing and non-dividing cells and may incorporate its genome into that of the host cell. Moreover, AAV mostly stays as episomal; performing long and stable expression. These features make AAV a very attractive candidate for creating viral vectors for gene therapy. However, AAV can only bring up to 5kb which is considerably small compared to AAV's original capacity.

Furthermore, because of its potential use as a gene therapy vector, researchers have created an altered AAV called Self-complementary adeno-associated virus (scAAV). Whereas AAV packages a single strand of DNA and requires the process of second-strand synthesis, scAAV packages both strands which anneal together to form double stranded DNA. By skipping second strand synthesis scAAV allows for rapid expression in the cell. Otherwise, scAAV carries many characteristics of its AAV counterpart.

Challenges in Application

The choice of a viral vector to deliver genetic material to cells comes with some logistical problems. There are a limited number of viral vectors available for therapeutic use. Any of these few viral vectors can cause the body to develop an immune response if the vector is seen as a foreign invader. Once used, the viral vector cannot be effectively used in the patient again because it will be recognized by the body. If the vaccine or gene therapy fails in clinical trials, the virus can't be used again in the patient for a different vaccine or gene therapy in the future. Pre-existing immunity against the viral vector could also be present in the patient rendering the therapy ineffective for that patient. It is possible to counteract pre-existing immunity when using a viral vector for vaccination by priming with a non-viral DNA vaccine, but this method presents another expense and obstacle in the vaccine distribution process. Pre-existing immunity may also be challenged by increasing vaccine dose or changing the vaccination route.

Lentivirus

Lentivirus (*lente-*, Latin for "*slow*") is a genus of retroviruses that cause chronic and deadly diseases characterized by long incubation periods, in man and other mammalian species. The best known lentivirus is the Human Immunodeficiency Virus HIV, which causes AIDS. Lentiviruses are also hosted in apes, cows, goats, horses, cats, and sheep. Recently, lentiviruses have been found in monkeys, lemurs, Malayan flying lemur (not a true lemur nor a primate), rabbits, and ferrets. Lentiviruses and their hosts have worldwide distribution. Lentiviruses can integrate a significant amount of viral RNA into the DNA of the host cell and can efficiently infect nondividing cells, so they are one of the most efficient methods of gene delivery. Lentiviruses can become endogenous (ERV), integrating their genome into the host germline genome, so that the virus is henceforth inherited by the host's descendants.

Classification

Five serogroups of lentiviruses are recognized, reflecting the vertebrate hosts with which they are associated (primates, sheep and goats, horses, domestic cats, and cattle). The primate lentiviruses are distinguished by the use of CD4 protein as a receptor and the absence of dUTPase. Some groups have cross-reactive *gag* antigens (e.g., the ovine, caprine, and feline lentiviruses). Antibodies to *gag* antigens in lions and other large felids indicate the existence of other as yet unidentified viruses related to feline lentivirus and the ovine/caprine lentiviruses.

Morphology

The virions are enveloped, slightly pleomorphic, spherical and measure 80–100 nm in diameter. Projections of envelope make the surface appear rough, or tiny spikes (about 8 nm) may be dispersed evenly over the surface. The nucleocapsids (cores) are isometric. The nucleoids are concentric and rod-shaped, or shaped like a truncated cone.

Genome Organization and Replication

Features of the genome: infectious viruses have three main genes coding for the viral proteins in the order: 5′-gag-pol-env-3′. There are two regulatory genes, tat and rev. There are additional accessory genes depending on the virus (e.g., for HIV-1: vif, vpr, vpu, nef) whose products are involved in regulation of synthesis and processing viral RNA and other replicative functions. The Long terminal repeat (LTR) is about 600 nt long, of which the U3 region is 450, the R sequence 100 and the U5 region some 70 nt long.

Viral proteins involved in early stages of replication include reverse transcriptase and integrase. Reverse transcriptase is the virally encoded RNA-dependent DNA polymerase. The enzyme uses the viral RNA genome as a template for the synthesis of a complementary DNA copy. Reverse transcriptase also has RNaseH activity for destruction of the RNA-template. Integrase binds both the viral cDNA generated by reverse transcriptase and the host DNA. Integrase processes the LTR before inserting the viral genome into the host DNA. Tat acts as a trans-activator during transcription to enhance initiation and elongation. The Rev responsive element acts post-transcriptionally, regulating mRNA splicing and transport to the cytoplasm.

Proteome

The lentiviral proteome consists of five major structural proteins and 3-4 non-structural proteins (3 in the primate lentiviruses).

Structural proteins listed by size:

1. Gp120 surface envelope protein SU, encoded by the viral gene *env*. 120000 Da (Daltons).
2. Gp41 transmembrane envelope protein TM, also encoded by the viral gene *env*. 41000 Da.
3. P24 capsid protein CA, encoded by the viral gene *gag*. 24000 Da.
4. P17 matrix protein MA, also encoded by *gag*. 17000 Da.
5. P9 capsid protein NC, also encoded by *gag*. 7000-11000 Da.

- The envelope proteins SU and TM are glycosylated in at least some lentiviruses (HIV, SIV), if not all of them. Glycosylation seems to play a structural role in the concealment and variation of antigenic sites necessary for the host to mount an immune system response.

Enzymes:

1. Reverse transcriptase RT encoded by the *pol* gene. Protein size 66000 Da.

2. Integrase IN also encoded by the *pol* gene. Protein size 32000 Da.
3. Protease PR encoded by the *pro* gene.
4. dUTPase DU encoded by the *pro* gene (part of *pol* gene in some viruses), the role of which is still unknown. Protein size 14000 Da.

Gene regulatory proteins:

1. Tat
2. Rev

Accessory proteins:

1. Nef
2. Vpr
3. Vif
4. Vpu/Vpx
5. p6

Antigenic Properties

Serological Relationships: Antigen determinants are type-specific and group-specific. Antigen determinants that possess type-specific reactivity are found on the envelope. Antigen determinants that possess type-specific reactivity and are involved in antibody mediated neutralization are found on the glycoproteins. Cross-reactivity has been found among some species of the same serotype, but not between members of different genera. Classification of members of this taxon is infrequently based on their antigenic properties.

Epidemiology

- Symptoms and host range: The virus' hosts are found in the orders Primates (humans, apes and monkeys), Carnivora (cats, dogs and other carnivores), Perissodactyla and Artiodactyla (single and double-toed hooved mammals).
- Transmission: Transmitted by means not involving a vector.
- Geographic distribution: Worldwide.

Physicochemical and Physical Properties

- General

- o Buoyant density 1.16–1.18 g cm^{-3} in sucrose
- o Virions sensitive to heat, detergents, and formaldehyde
- o Infectivity not affected by UV irradiation

Classed as having *class C* morphology

- Nucleic Acid
  - o Virions contain 2% nucleic acid
  - o Genome consists of a dimer
  - o Virions contain one molecule of (each) linear positive-sense single stranded RNA.
  - o Total genome length is of one monomer 9200 nt
  - o Genome sequence has terminal repeated sequences; long terminal repeats (LTR) (of about 600 nt
  - o The 5' end of the genome has a cap
  - o Cap sequence of type 1 m7G5ppp5'GmpNp
  - o 3' end of each monomer has a poly (A) tract; 3'-terminus has a tRNA-like structure (and accepts lysin)
  - o Encapsidated nucleic acid are solely genomic
  - o 2 copies packed per particle (held together by hydrogen bonds to form a dimer).
- There are 11 proteins
  - o Virions contain 60% protein
  - o Five (major)structural virion proteins have been found so far
- Lipids: Virions contain 35% lipid.
- Carbohydrates: Other compounds detected in the particles 3% carbohydrates.

Use as Gene Delivery Vectors

Lentivirus is primarily a research tool used to introduce a gene product into in vitro systems or animal models. Large-scale collaborative efforts are underway to use lentiviruses to block the expression of a specific gene using RNA interference technology in high-throughput formats. The expression of short-hairpin RNA (shRNA) reduces the expression of a specific gene, thus allowing researchers to examine the necessity and effects of a given gene in a model system. These studies can be a precursor to the development of novel drugs which aim to block a gene-product to treat diseases.

Another common application is to use a lentivirus to introduce a new gene into human or animal cells. For example, a model of mouse hemophilia is corrected by expressing wild-type platelet-factor VIII, the gene that is mutated in human hemophilia. Lentiviral infection has advantages over other gene-therapy methods including high-efficiency infection of dividing and non-dividing cells, long-term stable expression of a transgene, and low immunogenicity. Lentiviruses have also been successfully used for transfection of diabetic mice with the gene encoding PDGF (platelet-derived growth factor), a therapy being considered for use in humans. These treatments, like most current gene therapy experiments, show promise but are yet to be established as safe and effective in controlled human studies. Gammaretroviral and lentiviral vectors have so far been used in more than 300 clinical trials, addressing treatment options for various diseases.

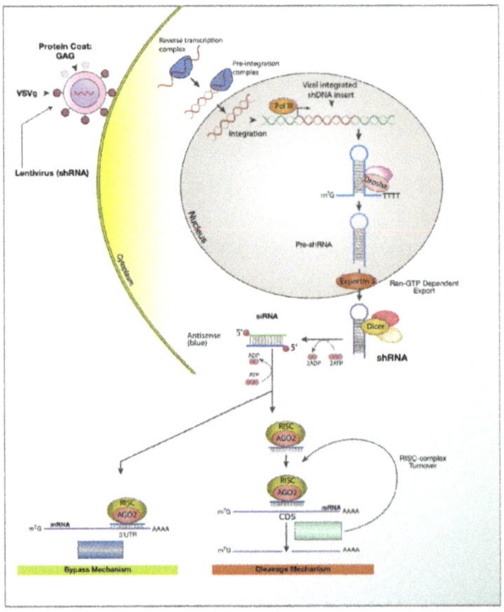

Lentiviral Delivery of designed shRNA's and the mechanism of RNA interference in mammalian cells.

Lentiviruses are enveloped RNA viruses found in most of the vertebrates. The most common Lentiviruses used in gene therapy experiments are retroviruses. The virion contains two copies of RNA genome, which are covered by a cone shaped core. The viral genome contains three genes that code for different viral proteins. The *gag* gene encodes for capsid, matrix, and nucleocapsid protein. The *pol* gene is a viral polymerase which comprises of proteases, reverse transcriptase and integrase. The *env* gene encodes for glycosylated envelope protein which mediates the virus entry into the host cell receptor. The viral RNA genome is flanked by long terminal repeats (LTR) sequences which are responsible for packaging of viral RNA genome. Most of the retroviruses lead to persistent infection by integrating to the host cell genome. Lentiviruses also contain two additional proteins *tat* and *rev* which take part in transactivation of viral transcription and nuclear export of viral RNA, respectively.

Following the infection virion binds to the host cell receptor and viral genome enters the cell by fusion. Viral RNA is converted into double stranded DNA with the help of enzyme reverse transcriptase. The double stranded DNA then migrates to the nucleus and integrates to the host cell genome by the help of enzyme integrase. This stage of the virus is known as Provirus . The proviral DNA forms new viral genome by using cellular transcription factors. The immature viral proteins are processed by viral proteases, and assemble along with RNA genome to form an infectious virion, which then buds out from the host cell membrane.

Most of the work based on the lentiviral vector focuses on modifying the human immune deficiency virus type 1 (HIV-1). HIV-1 encodes six accessory proteins namely *tat, rev, vif, vpr, nef,* and *vpu* in addition to an essential *gag, pol* , and *env* proteins.

Table: Different genes of lentivirus and their functions:

Gene	Name	Function
gag	Group specific antigen	Nucleocapsid core protein
pol	Polymerase	Encode reverse transcriptase, integrase, protease
vif	Viral infectivity fector	Virus infectivity
vpr	Viral protein R	Nuclear targeting
vpu	Viral protein unique	Help in virus budding
env	Envelope	Codes for surface coat protein
tat	Transactivator of transcription	Virus gene expression
rev	Regulator of expression of virion proteins	Structural gene expression
nef	Negative regulatory factor	Virus infectivity

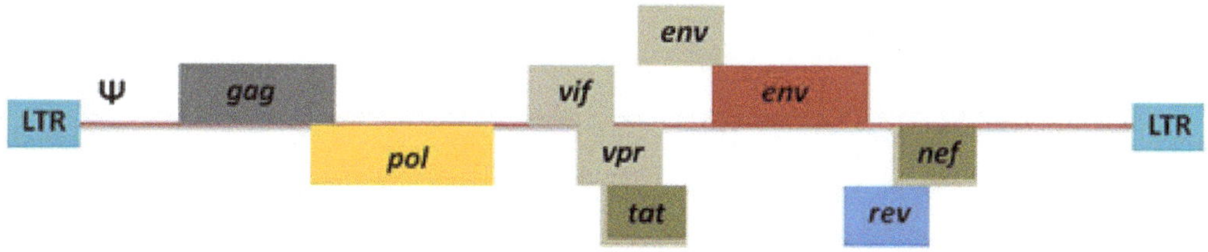

Schematic diagram of lentivirus

Gene therapy vectors based on lentiviruses lack all the viral sequences except the LTRs, *rev* responsive element (RRE), and *cis* -acting elements. Usually viral *rev* proteins are added in *trans* in order to facilitate the trafficking of viral RNA genome in the nucleus following efficient binding with RRE. The production of vector RNA is either directed by LTRs or by tissue specific promoter. Generally vectors are designed in such a way so as to utilize tissue specific promoter (CMV) and LTR hybrid promoter (CMV/LTR). The use of CMV/LTR hybrid promoter allows abundant vector RNA production independent of transactivation by *tat* protein. HIV accessory proteins, *vif, vpr, nef,* and *vpu* can be deleted during the lentiviral production. Sometimes polypurine tract present in the HIV genome can be included as a *cis* - active element during the viral production in order to enhance the nuclear trafficking. The HIV-1 infects only the cells that contain CD4 receptor and co-receptors such as CXCR4 and CCR5.The attachment of HIV to the receptor and co-receptors is largely mediated by viral glycoproteins. This specificity restricts the host range for HIV-1 infection. Scientists worldwide are trying to broaden the host range by various molecular biology techniques. One such technique is to generate a pseudo type HIV virus that contains vesicular stomatitis glycoprotein along with *env* protein, which is having broad tropism.

Diagram depicts a HIV based vector system where viral essential genes have been replaced by the transgene. Cytomegalovirus promoter is used to transcribe the transgene as well as gag, RRE, and cPPT. RRE acts a *cis* -acting sequence essential for nuclear export of viral RNA while polypurine tract (cPPT) helps in the import of proviral DNA. The deletion in U3 region in the genome makes it inaccessible to use LTR for transcription. The packaging system consists of *gag / pol* , VSV-G, and *rev* expressing constructs. Presence of RRE along with *gag / pol* helps in nuclear export of the RNA by transcribing *rev* protein. Lentiviral vectors are produced by transfection of vector construct along with the packaging constructs in producer cells. Vector RNA genomes are packaged by precursor proteins at the cellular membrane. The mature particles bud through the cellular membrane, containing the envelope derived from VSV-G glycoproteins.

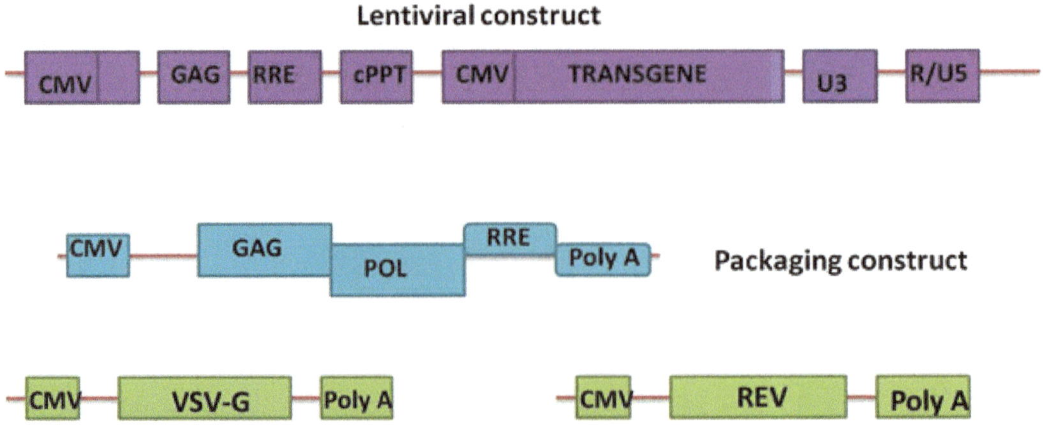

Generation of lentiviral vector

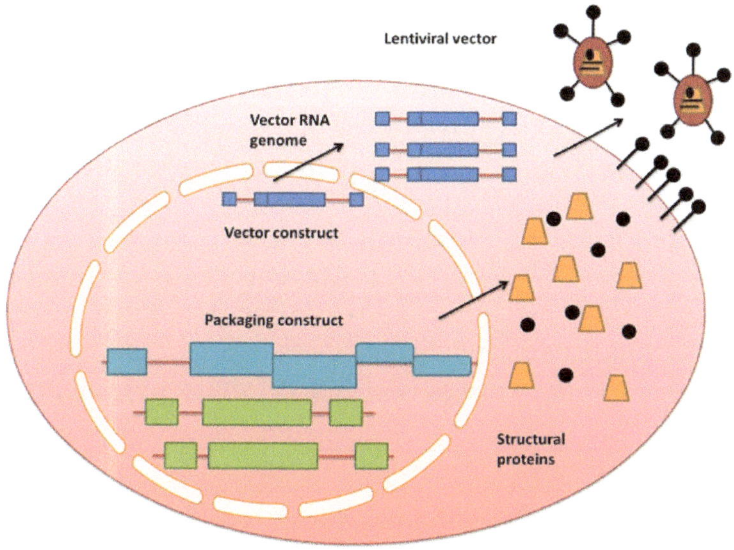

SV40

SV40 is an abbreviation for Simian vacuolating virus 40 or *Simian virus 40*, a polyomavirus that is found in both monkeys and humans. It was named for the effect it produced on infected green monkey cells, which developed an unusual number of vacuoles. Like other polyomaviruses, SV40 is a DNA virus that has the potential to cause tumors in animals, but most often persists as a latent infection.

The discovery of SV40 revealed that between 1955 and 1963 around 90% of children and 60% of adults in USA were inoculated with SV40-contaminated polio vaccines.

Historical Background

SV40 was first identified by Ben Sweet and Maurice Hilleman in 1960 when they found that between 10-30% of polio vaccines in the USA were contaminated with SV40. In 1962, Bernice Eddy described the SV40 oncogenic function inducing sarcoma and ependymomas in hamsters inoculated with monkeys cells infected with SV40. The complete viral genome was sequenced by Fiers and his team at the University of Ghent (Belgium) in 1978.

Virology

SV40 consists of an unenveloped icosahedral virion with a closed circular dsDNA genome of 5.2 kb. The virion adheres to cell surface receptors of MHC class I by the virion glycoprotein VP1. Penetration into the cell is through a caveolin vesicle. Inside the cell nucleus, the cellular RNA polymerase II acts to promote early gene expression. This results in an mRNA that is spliced into two segments. The small and large T antigens result from this. The large T antigen has two functions: 5% goes to the plasma cell membrane and 95% returns to the nucleus. Once in the nucleus the large T antigen binds three viral DNA sites, I, II and III. Binding of sites I and II autoregulates early RNA synthesis. Binding to site II takes place in each cell cycle. Binding site I initiates DNA replication at the origin of replication. Early transcription gives two spliced RNAs that are both 19s. Late transcription gives both a longer 16s, which synthesizes the major viral capsid protein VP1; and the smaller 19s, which gives VP2 and VP3 through leaky scanning. All of the proteins, besides the 5% of large T, return to the nucleus because assembly of the viral particle happens there. Eventual release of the viral particles is cytolytic and results in cell death.

Multiplicity Reactivation

SV40 is capable of multiplicity reactivation (MR). MR is the process by which two or more virus genomes containing otherwise lethal damage interact within an infected cell to form a viable virus genome. Yamamato and Shimojo observed MR when SV40 virions were irradiated with UV light and allowed to undergo multiple infection of host cells. Hall studied MR when SV 40 virions were exposed to the DNA crosslinking agent 4, 5', 8-trimethylpsoralen. Under conditions in which only a single virus particle entered each host cell, approximately one DNA cross-link was lethal to the virus and could not be repaired. In contrast, when multiple viral genomes infected a host cell, psoralen-induced DNA cross-links were repaired; that is, MR occurred. Hall suggested that the virions with cross-linked DNA were repaired by recombinational repair. Michod et al. reviewed numerous examples of MR in different viruses and suggested that MR is a common form of sexual interaction that provides the advantage of recombinational repair of genome damages.

Transcription

The early promoter for SV40 contains three elements. The TATA box is located approximately 20 base-pairs upstream from the transcriptional start site. The 21 base-pair repeats contain six GC boxes and are the site that determines the direction of transcription. Also, the 72 base-pair repeats are transcriptional enhancers. When the SP1 protein interacts with the 21 bp repeats it binds either the first or the last three GC boxes. Binding the first three initiates early expression and binding the last three initiates late expression. The function of the 72 bp repeats is to enhance the amount of stable RNA and increase the rate of synthesis. This is done by binding (dimerization) with the AP1 (activator protein 1) to give a primary transcript that is 3' polyadenylated and 5' capped.

SV40 in Animals

SV40 is dormant and is asymptomatic in rhesus monkeys. The virus has been found in many macaque populations in the wild, where it rarely causes disease. However, in monkeys that are immunodeficient—due to, for example, infection with *Simian immunodeficiency virus*—SV40 acts much like the human JC and BK polyomaviruses, producing kidney disease and sometimes a demyelinating disease similar to PML. In other species, particularly hamsters, SV40 causes a variety of tumors, generally sarcomas. In rats, the oncogenic SV40 large T-antigen was used to establish a brain tumor model for PNETs and medulloblastomas.

The molecular mechanisms by which the virus reproduces and alters cell function were previously unknown, and research into SV40 vastly increased biologists' understanding of gene expression and the regulation of cell growth.

Hypothesized Role in Human Disease

The hypothesis that SV40 might cause cancer in humans has been a particularly controversial area of research. Several methods have detected SV40 in a variety of human cancers, although how reliable these detection methods are, and whether SV40 has any role in causing these tumors, remains unclear. As a result of these uncertainties, academic opinion remains divided, with some arguing that this hypothesis is not supported by the data and others arguing that some cancers may involve SV40. The US National Cancer Institute announced in 2004 that although SV40 does cause cancer in some animal models, "substantial epidemiological evidence has accumulated to indicate that SV40 likely does not cause cancer in humans". This announcement was based on two studies. This 2004 announcement is in contrast to a 2002 study performed by The National Academy of Sciences Immunization Safety Review committee that stated, "The committee concludes that the biological evidence is moderate that SV40 exposure could lead to cancer in humans under natural conditions." However, Namika, Goodison,...and Rosser found that the SV40 large t-antigen, in combination with mycoplasma, often a contaminate of vaccines, can cause prostate cells to turn cancerous. Whether or not this is true for other human cells is debated.

p53 Damage and Carcinogenicity

SV40 is believed to suppress the transcriptional properties of the tumor-suppressing p53 in humans through the SV40 large T-antigen and SV40 small T-antigen. p53 is responsible for initiating regulated cell death ("apoptosis"), or cell cycle arrest when a cell is damaged. A mutated p53 gene may contribute to uncontrolled cellular proliferation, leading to a tumor.

SV40 may act as a co-carcinogen with crocidolite asbestos to cause both Peritoneal and Pleural Mesothelioma.

When SV40 infects nonpermissive cells, such as 3T3 mouse cells, the dsDNA of SV40 becomes covalently integrated. In nonpermissive cells only early gene expression occurs and this leads to transformation, or oncogenesis. The nonpermissive host needs the large T-antigen and the small t-antigen in order to function. The small T-antigen interacts with and integrates with the cellular phosphatase pp2A. This causes the cell to lose the ability to initiate transcription.

Polio Vaccine Contamination

Soon after its discovery, SV40 was identified in the oral form of the polio vaccine produced between 1955 and 1961 by American Home Products (dba Lederle). This is believed to be due to two sources: 1) SV40 contamination of the original seed strain (coded SOM); 2) contamination of the substrate—primary kidney cells from infected monkeys used to grow the vaccine virus during production. Both the Sabin vaccine (oral, live virus) and the Salk vaccine (injectable, killed virus) were affected; the technique used to inactivate the polio virus in the Salk vaccine, by means of formaldehyde, did not reliably kill SV40. The contaminated vaccine continued to be distributed to the public through 1963.

It was difficult to detect small quantities of virus until the advent of PCR; since then, stored samples of vaccine made after 1962 have tested negative for SV40. In 1997, Herbert Ratner of Oak Park, Illinois, gave some vials of 1954 Salk vaccine to researcher Michele Carbone. Ratner, the Health Commissioner of Oak Park at the time the Salk vaccine was introduced, had kept these vials of vaccine in a refrigerator for over forty years. Upon testing this vaccine, Carbone discovered that it contained not only the SV40 strain already known to have been in the Salk vaccine (containing two 72-bp enhancers) but also the same slow-growing SV40 strain currently found in some malignant tumors and lymphomas (containing one 72-bp enhancers). It is unknown how widespread the virus was among humans before the 1950s, though one study found that 12% of a sample of German medical students in 1952 had SV40 antibodies.

An analysis presented at the Vaccine Cell Substrate Conference in 2004 suggested that vaccines used in the former Soviet bloc countries, China, Japan, and Africa, could have been contaminated up to 1980, meaning that hundreds of millions more could have been exposed to the virus unknowingly.

Population level studies show no evidence of any increase in cancer incidence as a result of exposure, though SV40 has been extensively studied. A thirty-five year followup found no excess of the cancers putatively associated with SV40.

Simian virus 40 (SV40) belongs to polyoma viruses group. The virus was first isolated as a contaminant from a monkey kidney culture used for the production of poliovirus vaccine. Its genome is very small and can be easily modified for gene therapy purpose.

Its genome has ds circular DNA wrapped up as nucleosome with the help of histone proteins (H1, H2 (A and B), H3, H4). It is roughly 40-45 nm in diameter. The chromosome is 5000 bp long. The genome of SV 40 is like a minichromosome. The genome of SV40 consists of early proteins, late proteins and regulatory proteins. Early proteins are non structural while late proteins are structural proteins. Regulatory region consist of promoter, enhancer and origin of replication. The virions are made up of three capsid proteins namely VP1, VP2, and VP3. MHC class I molecules act as receptors for SV 40 virus. Following attachment to the cell surface by VP1 virus gets internalized by endocytosis. The viral genome replication takes place inside the nucleus. The host RNA polymerase II helps in the transcription of early gene products. The early proteins produced by the virus once it enters the cell consist of T antigen, t antigen and middle T antigen. The large T antigen migrates to the cytoplasm while majority of it gets back to the nucleus during virus formation before release. The capsid proteins are produced later which then forms viable virus and comes out of the cell. The early protein coding genes can be replaced by the gene of interest. The mutation in p53 promoter which leads to cancerous condition was discovered in the SV 40 virus.

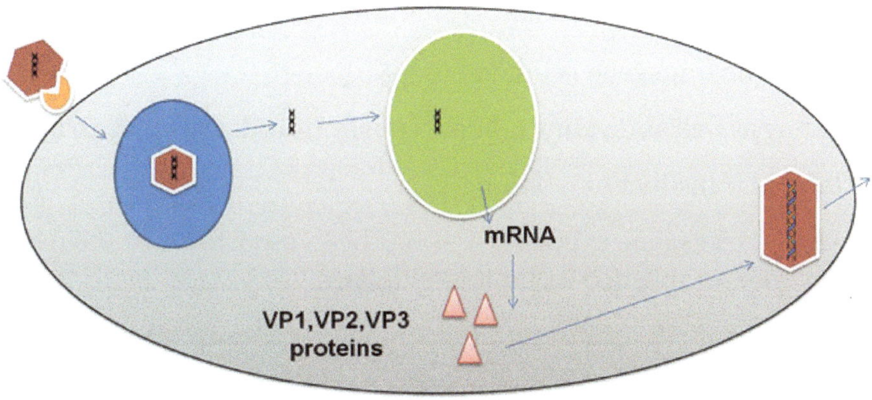

Replication of SV40 in the host cell:

SV40 as a Vector

Gene therapy using recombinant SV40 is highly suitable means to transfer the gene in a number of conditions. The wild type SV40 virus encodes two early genes, a large T antigen and a small t antigen. The large T antigens are important for virus replication and synthesis of viral late proteins. The genome of SV40 can be modified in order to carry foreign gene by deleting large T antigen (approximately 2.5 kb). The most alarming situation in a wild type SV40 is regarding the ability of large T antigen to bind with tumor suppressor genes such as p53 and retinoblastoma (pRb) protein. The binding of large T antigen may lead to cancer because of inactive p53 and pRb proteins. To avoid this complication all SV40 recombinant viruses lack large T antigen.

The late proteins of SV40 are synthesized from the opposite strand. Out of four structural proteins VP1 is the most abundant one. The minimum viral specific sequence required to assemble the virion consists of origin of replication and encapsidation sequences. The origin of replication and encapsidation sequences overlaps with the early promoter region in the genome. The transcription of the genome and the transgene is governed by pol III promoter. The transcription of the transgene is terminated by the incorporation of a polyadenylation signal downstream to the gene of interest. Most of the vectors available for the expression of transgene requires SV40 early and late polyadenylation signal.

Recombinant SV40 viral vectors are made in some plasmid background such as pGEMT plasmid vector. The permissible cells are transfected with the plasmid along with the support plasmid needed to package the viral genome. The recombinant virus containing the transgene is then recovered by the lysis of transfected cell. The recombinant virus is then further amplified in the next round of infection in the cell lines. This approach is used to package the transgene of about 5.5 kb without affecting the productivity.

Features of a SV40 Based Vector

Some of the specific features of an SV40 vector are as follows

- Capacity of the transgene is limited to 5 Kb.
- Both resting as well as dividing cells are equally transduced by SV40 based vectors.
- High level of transduction efficiency.
- Long lasting transgene expression. It has been reported that the transgene expression once established will remain lifelong.
- Neutralizing antibody against recombinant SV40 has not been reported yet. Therefore it is easy to administer multiple injection of recombinant SV40 vector without neutralizing its effect.
- SV40 readily integrates with the host cell genome both in dividing and non dividing cells.
- The recombinant SV40 is quite stable and can be stored in lyophilized form.
- Protein production from a transgene is optimum (not high like adenovirus).
- Production titer of recombinant SV40 is very high so a large quantity of cells can be transduced by the vector.

Many cell types have been tested successfully for the transgene expression by the recombinant SV40, such as hepatocyte, lymphocytes, kidney cells, brain cells, skin, and lungs etc. However, the cell types such as cardiac and smooth muscles of blood vessels are known to transduce less efficiently by SV40 vectors.

Safety Issue with Recombinant SV40 Based Vectors

The major issue of using recombinant SV40 vectors is contamination of wild type SV40 in the recombinant population and insertional mutagenesis. The first issue is largely related with the cell line used for the production of recombinant virus. Mostly COS7 cells are used to produce the recombinant SV40 because of the high yield of the virus. However, COS7 cells are reported to produce wild type population during the subsequent passage. The appearance of wild type population in the samples can be avoided by using packaging cell lines expressing the viral structural proteins designed in such a way as not to overlap with the wild type sequences.

Wild type SV40 virus is known to integrate with the cellular genome for a long time. Although the site of the integration of SV40 is not fixed and varies each time it infects the host cell. It is quite obvious that the activity of the genes changes because of the integration of any foreign DNA fragments. The insertion of foreign gene may activate or inactivate the gene. The activity of gene becomes down-regulated if the insertion occurs in between the open reading frame of the gene. The insertion of the foreign DNA may lead to death of the cells if the function of the affected gene is critical for the cell cycle. The most bothering part about the integration is regarding the activation of a cellular gene because of the activation of promoter. The activation is more alarming when the activation is of cellular oncogene. The effect is more pronounced in retroviral vector based vectors.

Non Viral Vectors

Traditional Non Viral Methods of Transgenesis

With the advent of using exogenous genetic element for treating diseases various methods for delivering transgene have been developed. These conventional methods can be broadly classified into physical and cationic polymer methods.

Physical Methods of Transgenesis

a. Hydrodynamic pressure techniques : In this technique intravascular injection of plasmid DNA is given to drive DNA molecules out of the blood circulation and into the tissue. This method of transgenesis is limited to locations which can tolerate temporary increase in pressure.

b. Electroporation : In this method controlled electric shocks are given to cells resulting in formation of large cytoplasmic pores, through which polynucleotide can move into the cell. Such methods are mostly restricted to *in vitro* applications but development of new electrodes designed for *in vivo* applications has allowed its use in certain areas of the body, such as tumors, muscles and liver tissue in animal models .

c. Ballistic delivery : In ballistic delivery or particle bombardment, DNA-coated metal microparticles are allowed to penetrate cell membranes at high velocity. *In vivo* this technique is limited to cutaneous applications.

d. Microinjection : In this technique DNA is injected into cells resulting in efficient transgene expression but is laborious and limited to *ex vivo* applications, such as for the delivery of artificial chromosomes.

Cationic Polymer Method of Transgenesis

a. Lipofection : This method involves the use of cationic lipids mixed with DNA in aqueous solution forming liposomes that encapsulate DNA. Such DNA preparations are taken in either by pinocytosis or phagocytosis, depending on the cell type. Use of liposomes *in vivo* is limited by its rapid plasma clearance and high toxicity.

b. Cationic peptides : These are chain of basic amino acids, which compact DNA into spherical complexes, or chromatin components such as histones or protamine, which compact DNA in a structured manner allowing it to enter cell by their interaction with sulfated membrane-bound proteoglycans.

c. Polyethylenimine : Polyethylenimine is a linear or branched polymer with many protonable amino nitrogen atoms that allows efficient DNA condensation and cell entry. The polymer behaves as an effective 'proton sponge' that causes the rupture of endosome by osmotic swelling and releases the polyethylenimine–DNA complexes into the cytoplasm.

d. Receptor-mediated delivery : In this method polycations conjugated DNA bound to a cell-specific ligand is targeted to cells via receptor-mediated internalization.

All these non viral vectors have proven well in *in vitro* transfection but lacks in *in vivo* transfection efficiency and only allow transient transgene expression. This inefficiency of non viral vectors can be attributed to the following:

- Interaction of the non viral–DNA complex with blood plasma proteins, undesirable cells and extracellular matrix.
- Inability to escape from liposome or endosome enclosed moiety.
- Vulnerability to cytoplasmic degradative enzymes.
- Inability to pass through the double membrane nuclear envelope and subsequent degradation during breakdown of nuclear membrane at mitosis.

Thus to increase the *in vivo* stability of non viral vectors novel modular vectors have been developed. These vectors have reduced affinity for intracellular proteins and cell surfaces and posses ligands for receptor-mediated endocytosis, peptide sequences that assist DNA compaction, endosomal disruption sequences and nuclear-import signals. Modular vectors can be easily modulated and they also mimic the ability of viruses to overcome the cellular barriers to DNA delivery. GD5 is one such modular vectors, which forms complex with DNA using DNA binding domain (DBD) of GAL4 transcription factor, attaches to tumor cells via a single chain antibody against ERBB2 antigen leading to receptor mediated endocytosis and facilitates endosomal escape using translocation domain of diphtheria toxin in acidic pH.

Modular vectors can provide efficient delivery of therapeutic gene complexed to non viral vectors but stable nuclear maintenance and replication of such vectors can be achieved by either site specific integration into safe genomic regions or maintenance of extra chromosomal segments.

Safe Integration of Corrective Genes to Human Genome

The problem of insertional mutagenesis in gene therapy can be avoided by directing transgene into safe genomic locations that are not associated with cell proliferation or tumor suppression. Site specific recombination used by prokaryotic viruses may be helpful in this regard. Prokaryotic viruses express site specific recombinases (SSRs) that recognize unique genomic sequences called recombination sites within pathogen and host genome and mediate recombination at those sites. SSRs recognize defined but degenerate sequences and thus can be manipulated for integration into eukaryotic genome. Thus combination of SSR and non viral vectors can be useful in making safe integrations resulting in permanent expression of the therapeutic gene. For proper integration SSR must recognize specific recombination sites, for example *loxP* in case of bacteriophage *Cre* recombinase. Thus for use in human gene therapy these SSRs must be able to recognize endogenous sites in human genome called pseudosites that resemble phage recombination sites. Mammalian genomes contain several pseudo sites, e.g, bacteriophage *Cre* recombinase is capable of integrating between phage *loxP* site and pseudo *loxP* site in human genome albeit with 4 fold lower efficiency.

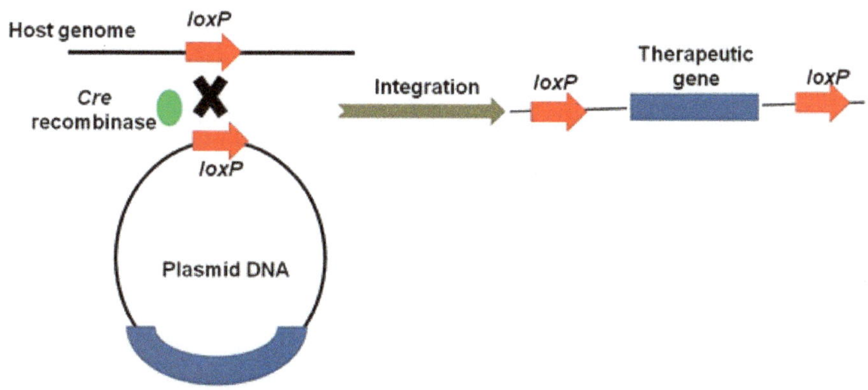

Site specific recombination by ***Cre*** recombinase

SSRs from phages (*Cre*) and from yeast (*Flp*) catalyses both excision and integration in equilibrium and because the excision reaction is favored this results in inefficient integration and subsequent expression. Other SSR like phage γ integrase create hybrid recombination sites and require separate excisionase for its function. Some phage SSR like those from λ, HXO22 require cofactors like IHF for recombination whereas other SSR like those from phage R4, TP901 and ΦC31 donot require cofactors and are capable of stable integration.

ΦC31 SSR is the most efficient non mammalian SSR for human gene therapy. This recombinase recognizes pseudosite psA in human genome and mspA and mspL1 in mice and is capable of treating recessive genetic disorders. ΦC31 is capable of integrating large gene like alpha typeVII collagen (COL7A1) and dystropin gene to be used in the treatment of RDEB and Duchenne muscular dystrophy .

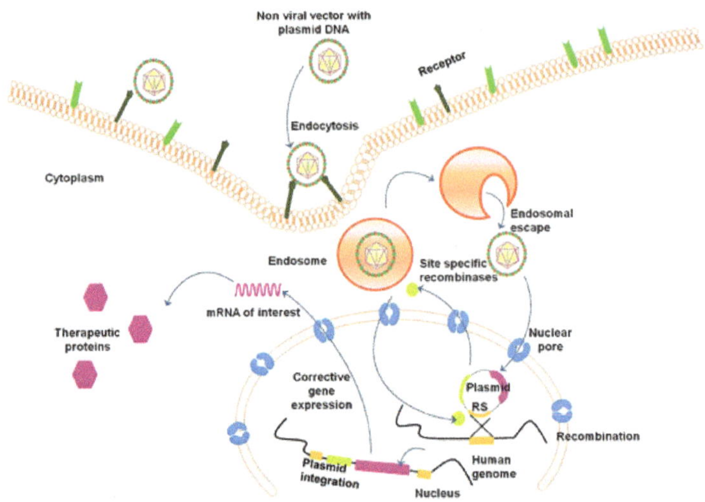

Mechanism of integration of therapeutic gene in human genome using site specific recombinases:

Prokaryotic origin of SSRs limits their application in eukaryotic hosts. Efficient transgene expression might be improved by using exogenous nuclear localization signal. Moreover, directed evolution is another approach to improve transgenesis by prokaryotic SSRs. In this technique the SSRs are subjected to multiple rounds of random mutagenesis and screening. *Flpe*, a derivative of *Flp* recombinase from yeast was developed using directed evolution. Various high throughput methods have now been developed to improve SSR activity that includes FACS and PCR based methods.

An attractive alternative of prokaryotic SSRs is Adeno Associated Virus (AAV). AAVs are non pathogenic mammalian virus and used viral encoded Rep protein to target transgene into specific sites called AAVS1, present on chromosome 19 in humans. The rep and cap genes in wild type AAV are bounded by 145 bp inverted terminal repeats (ITR). Rep is a replication protein with auxiliary recombinase function. Of the five alternatively spliced rep products only Rep68 and Rep78 catalyze site specific recombination by forming complexes with AAV ITRs and AAVS1 locus, followed by DNA strand cleavage and non homologous recombination.

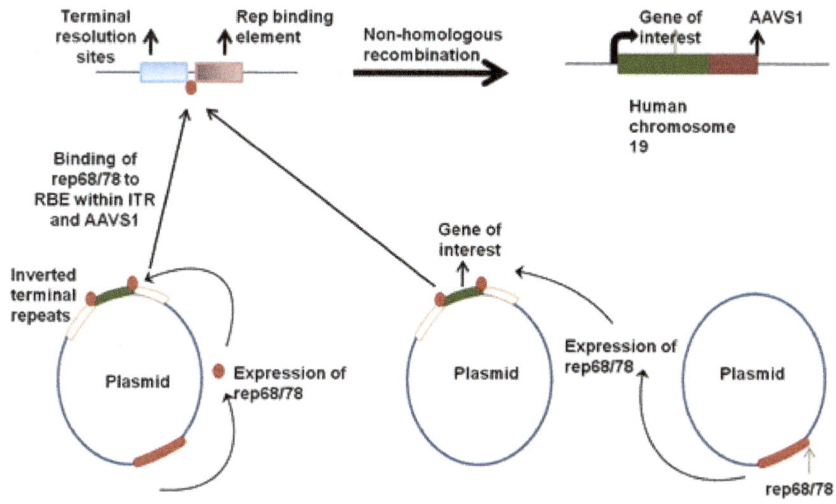

Rep protein mediated site specific recombination in Adeno-associated virus

The use of AAV is limited by certain level of immunogenicity it induces in humans and restriction in packaging 5 kb gene fragment. Recombinant AAV can be efficiently used in human gene therapy where the rep and cap genes are replaced by therapeutic genes. Plasmid containing Rep68 and Rep78, delivered by non-viral vectors along with ITR flanked transgene can efficiently integrate the corrective gene in AASV1 site in human genome. Constitutive expression of Rep is cytotoxic and cytostatic and sometimes results in abortive rearrangements of AASV1. To overcome this problem Rep can be provided in *trans* , from a cotransfected plasmid that lacks the Rep binding element. Alternatively the expression of Rep in host cells can be modulated by using ligands capable of inducing Rep or the protein can be externally delivered along with the plasmid to catalyze single integration event.

Retroviruses are another major source of integrase for human gene therapy. Retroviral integrase can be fused with sequence specific DNA binding proteins, such as phage γ repressor, *E.coli* LexA repressor, the Ty3 tRNA binding domain or designed Zn finger protein E2C for use in gene therapy.

Extra Chromosomal Replicating Units in Transgenesis

DNA molecules that can be replicated and segregated to daughter cells in an autonomous, extrachromosomal form avoiding insertional mutagenesis can be developed as another method of transgenesis in gene therapy. DNA viruses such as SV40 and Epstein–Barr virus (EBV) normally replicate episomally in mammalian cells and transfer viral episomes to daughter cells during cell division. Episomal plasmid vectors with viral ori sequence and requisite viral proteins (like Tag and EPVNA1) have been generated that can sustain the expression of a therapeutic transgene. For example, EBV-based episomal plasmid with 115 kb hypoxanthine phosphoribosyl transferase (*HPRT*) locus that was stably expressed in *HPRT* -deficient human lung fibroblast cells. Unfortunately, use of such episomes in human gene therapy is restricted as viral proteins such as Tag, alter the function of other key host cell proteins interfering with the retinoblastoma and p53 tumour-suppressor pathways.

To overcome this problem 'pEPI' plasmid containing an SV40-ori sequence and scaffold/ matrix attachment region (S/MAR) from the human β-interferon gene cluster has been developed that can propagate episomally, for several hundred divisions in CHO cells and HeLa human cervical cancer cells. Mitotic stability in pEPI results from the interaction of the S/MAR element with components of the nuclear matrix which is crucial for the organization of 'boundary elements' to protect their coding regions and to allow transcription factors to access enhancer and/or promoter sequences. Interaction of pEPI is mediated by binding of S/MAR to scaffold attachment factor A proteins, allows co-segregation of pPEPI with mitotic chromosomes and brings it closer to host replication machinery facilitating its own replication along with the host cell cycle. Transcription and replication are linked in pPEPI. Transcription of the AT rich S/MAR sequence is important in opening of chromatin structure to allow replication of pEPI and thus replication of this plasmid is not solely dependent on the SV ori. This property of chromosomal components like S/MAR can thus be exploited to develop novel episomally replicating plasmids for sustained expression of therapeutic genes in gene therapy.

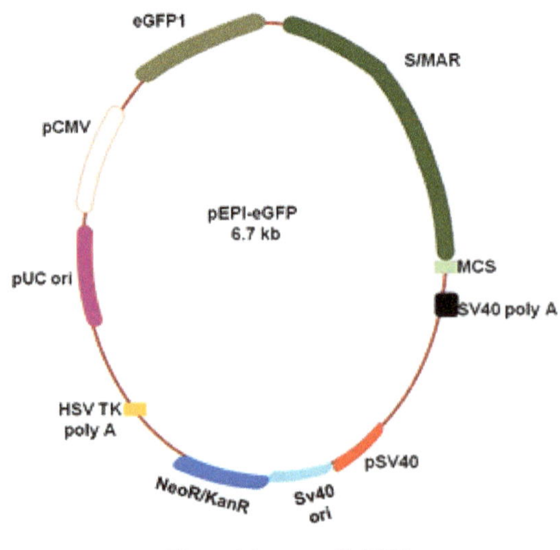

Plasmid map of pEPI

Human Artificial Chromosome

Artificial chromosomes are replica of normal chromosomes that are capable of replication and retention at low, defined copy number and within host cells. They have three main parts: a centromere, telomeres at both termini, and ori of replication. These chromosomes have exceptionally high cloning capacity and thus provide an excellent approach to be used as vectors in gene therapy. The first prototype HAC was generated in1997.

Human Artificial Chromosomes (HAC) can be synthesized *de novo* by assembly of the major components or by truncation of mammalian chromosome down to mitotically stable minichromosomes of 1-10 Mb, consisting of alpha satellite DNA arrays. Also neocentromeres based chromosomes provide another alternative of synthesizing HAC. Neocentromeres are functional variant of human centromeres and naturally present in non-centromeric region that are devoid of alpha satellite DNA. Neocentromere based human minichromosomes are 0.7-2 Mb in size that bind to centromere protein and are mitotically stable. HAC have been used in several gene therapy studies, for example in long term expression of 40 kb HPRT gene in Lesch-Nyhan syndrome, 250 kb CFTR gene and regulatory sequence in treating cystic fibriosis.

Because of its large size, microcell mediated chromosome transfer and microinjection are used to transfect HAC into different cell types in *ex vivo* approaches. Conventional non viral approaches make use of liposomes along with polyethylenimine or polylysine to deliver artificial DNA but are of lesser efficiency *in vivo* . The use of lipids with ultrasound has recently been demonstrated to induce better membrane pore formation allowing entry of DNA. Addition of NLS sequence either to the delivery agent or HAC might further improve nuclear entry or help stable transfection.

Maintenance of Transgene Expression in Host System

With evolution eukaryotic systems have developed mechanisms to maintain genome integrity and prevent foreign gene expression. Transient expression of transgenes introduced by viral or non viral vectors and episomal plasmids is in part due to *de novo* cytosine methylation by DNA methyltransferases that transcriptionally silent chromatin structures. Transgene expression is more favored by viral promoters as compared to human housekeeping promoters, but cytosines in viral promoters are main targets for cellular methyltransferases. Thus promoter development has been focused by mix-matching transcription factor binding sites for strong and long term expression. Also the addition of S/MAR sequence can provide position independent transcription and inhibit transgene methylation. In nonviral gene therapy the bacterial plasmid backbone used also contributes to transgene silencing. To overcome this problem phage recombination system have been used to form minicircle DNA that can be maintained as episomes *in vivo* and facilitate 560 fold higher transgene expression. mRNA splicing is another barrier to stable transgene expression that can be reduced by altered codon used to remove splice junctions. Moreover, immune response against accumulated foreign proteins in host circulation might restrict transgene stability which can be reduced by employing specific T cells, but further study is required in this aspect.

Future Prospects

Non viral procedures of gene therapy mimic the basic viral mechanisms to achieve long term expression of the corrective gene without disturbing host gene expression and signaling pathways. The efficiency of targeted delivery is the key to clinical success of non viral gene therapy. The effectiveness and difficulties of such delivery can now be well determined in terms of quantification, magnitude and duration of reporter gene expression, using small animal imaging technologies *in vivo* . Safe and long term transgene expression is the ultimate requirement of human gene therapy. Today gene therapy is not only restricted in proving for a missing gene but being explored in various aspects like anti-cancer gene therapy etc. and tailoring non-viral gene delivery strategies might help in realization of the potentials of human gene therapy.

DNA Vaccines

The earliest form of protection against the diseases has been in the form of vaccines. Discovery of vaccines by Edward Jenner in the year 1796 witnessed a revolution in human medicine. A vaccine is a biological formulation that confers immunity to a specific disease. The role of the vaccine is to stimulate the immune system, recognize the attacking agent, eradicate it and keep it in memory so that if there is any repeated exposure of the same disease then the destruction of such agent would be easier. There are several types of vaccines like killed, attenuated, live, subunit, conjugated and Nucleic acid vaccines.

Nucleic acid vaccine means the vaccination done using RNA or DNA vaccines. Here we are going to deal with DNA vaccines in full detail and RNA part will be summarized briefly. In a layman's words DNA vaccine can be defined as the process of injecting manipulated DNA material in an organism to generate an immune response. In general DNA vaccines may be defined as those vaccines that provide immunity by transfecting host cells with DNA that encodes an antigen. Following transfection the specific immune response generated is along similar lines as responses that occur against conventional vaccines and this immune response arises by the protein produced by the host cell. The DNA vaccine seems to be a promising candidate as vaccines because they are very cost effective but still these are not listed for human use. There are some DNA vaccines for veterinary use though.

The reasons for announcing DNA vaccine as potential candidate are many but also when compared to other forms of vaccine it proves to be less cumbersome and offers many advantages too enlisted below-

1) It is easy to form and construct the DNA vaccines as compared to attenuated viruses and subunit protein vaccines.

2) It is cheap and cost effective.

3) It is quite stable at room temperature as compared to attenuated viral vaccines, whose repository and international transfer becomes difficult by the urgency to keep the vaccines cold.

4) The protection offered by DNA vaccines favors bias towards cellular immunity, which is trusted to be vital for successful vaccination against intracellular pathogens.

The classic methods for synthesizing vaccines are given in table which highlights their similarities or differences with DNA vaccines.

Table: Comparison of different vaccine technologies:

Live attenuated viruses
Very effective
Potential risk for certain ones,
Manufacturing challenge
Viral vectors
Potential risk Resistance/pre-existing antibody,
Inflammation
Recombinant proteins
Effective antibody response, Effective, Non-native forms sometimes do not induce cytolytic T lymphocytes
DNA vaccines
Inevitable need for increased potency
Designer immune responses
Specificity: avoid deleterious or diversional antigens
Relative stability
Safety
Generic manufacturing
Cost advantage

Despite the fact that viral vectors and DNA vaccines have parallel attributes as given in table, they are only in their initial stages of clinical progress.Currently researchers worldwide are working on designing the vaccines in which the induction of CD8+ cytolytic T lymphocytes (CTL) responses and antibodies is emphasized due to the prominent role of CTL in such vaccines. Similarly designing of vaccines that can induce specific types of T helper responses, Th1 or Th2 are encouraged for the same reasons.

The diagram depicts the intracellular and intercellular interactions needed for the antigen to produce both cytotoxic and T helper cell responses as well as the production of antibodies. The general fate of a recombinant protein or an inactivated vaccine is that it is recognized by antigen presenting cells, then degraded into peptides and further allies with the major histocompatibility complex (MHC) class II molecules and this is the reason why such vaccines do not produce a required CTL response. In future these peptide or MHC complexes trigger T helper cells. For the production of CTL response, the viral infected cell must enter the cellular processing pathway from the cytoplasm so that the peptide gets associated with MHC class I molecules. These peptides subsequently are identified by specific cytolytic T cells and then are stimulated to kill the infected cell. Thus after successful delivery of a gene encoding an antigen into the cell, cytoplasm would be the site where the synthesized protein would be located and if such protein enters the intracellular processing pathway it can result in desired CTL response after getting associated with MHC class I molecules. Live vaccine virus is the best example to achieve this gene transfer with desired CTL response. Somehow this delivery method is not applicable to many virulent diseases as the live attenuated virus vaccines have the tendency to revert back to the wild type or they may successfully down-regulate the immune response. HIV live virus vaccine for example can revert back to the virulent type and thus can be more dreaded and fatal.

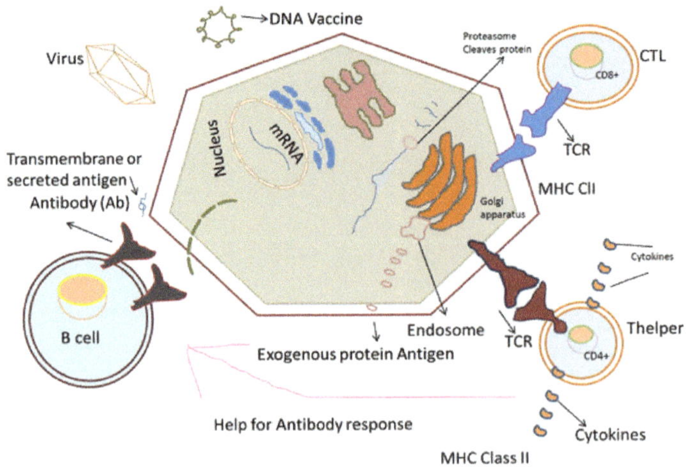

Mechanism of cytotoxic T cells, helper cells and antibodies

Features of DNA Vaccines

It is the efficient machinery of the virus that allows it to introduce its genetic material into the host cell. This is the reason why it was unbelievable for many researchers that a simple plasmid could enter the mammalian cell and synthesize the desired protein. Use of DNA plasmids as gene delivery vehicles is very convenient as compared to other methods as it does need any modification except a promoter active in mammalian cells. Still the use of DNA vaccines is a topic of debate because of its safety concerns.

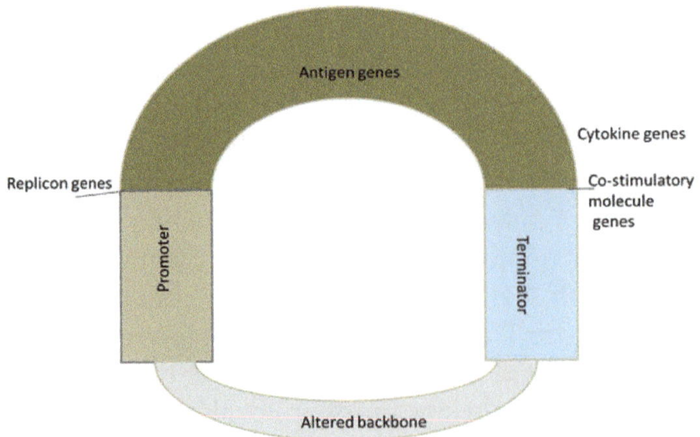

Schematic representation of a DNA vaccine

Following DNA vaccination communication between surfaces of the T-cell and the antigen presenting cells (Co-stimulatory molecules) as well as recognition of MHC class I complex by the T-cell receptor is must for a successful antigen presentation to a naïve T cell to stimulate the cell. Co-stimulatory molecules are absent in muscle cells and hence they do not behave like efficient antigen presenting cells. Usually if a naïve T-cell confronts a cell carrying a right antigen- MHC Class I complex in the absence of the co-stimulatory molecules or antigen presenting cells then the T-cells turn insensitive to this antigen if there is any repeated exposure to the same antigen.

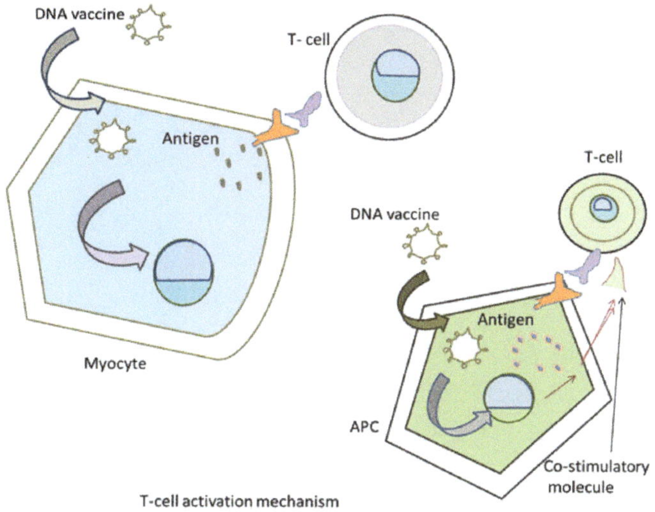

Antigen presentation after DNA vaccination

Trials with DNA Vaccines:

After several failures the first fairly successful DNA vaccination result in humans was attained with a malaria- specific DNA vaccine. This trial validated the rise of vaccine specific T cells in the peripheral blood of 11 out of 20 malaria-naive volunteers after three intramuscular pDNA (plasmid DNA) inoculations. However the investigation did not evaluate clinical gain. Several trials have offered evidence of principle for the ability of DNA vaccines to induce humoral and cellular immune responses in humans. Although, the immune responses analyzed were not as good as expected from the preclinical studies in any of these trials. For example, HIV-infected patients with huge viral counts generated a normal T-cell response against HIV Nef following DNA vaccination with a DNA vaccine encoding several HIV antigens, with no effect on viral counts.In an attempt to blend the qualities of DNA vaccines with those of adenoviral- or MVA-based vaccines, so-called 'heterologous prime-boost' systematic plans have been developed. In these plans the potent but broad immunity induced by MVA- or adenoviral- based vaccines gets focused on the appropriate antigens by DNA vaccination which is highly specific. Recently investigations based on DNA vaccines have shifted largely to tumor antigens and the reason behind it might be the excelling funding opportunities in this field.

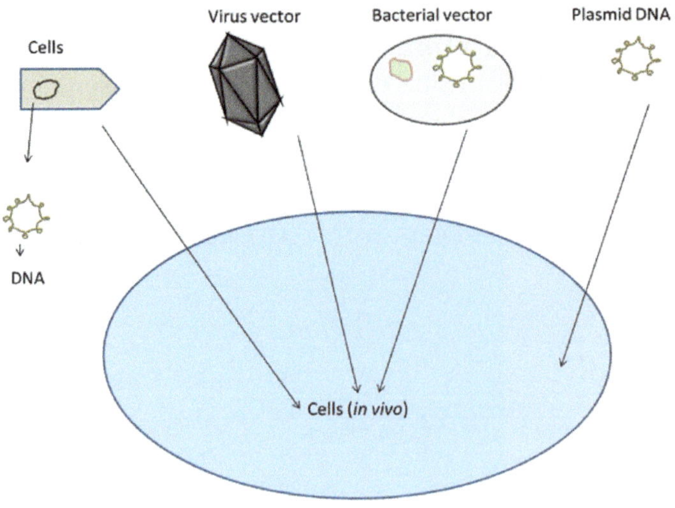

Techniques of delivering DNA vaccine

Due to discouraging results in the clinical trials, the DNA vaccination field is putting a lot of attention on upgrading the delivery methods, carrier molecules and genetic optimization of the construct used. Generally administration of DNA vaccines is done either by intramuscular (IM) or by intradermal (ID) route. Following IM route the antigen mainly is generated in muscle cells and this route promises high antigen expression but does not prove to be very immunogenic because muscle cells generally lack antigen presenting cells. On the contrary ID route has scanty antigen expression but it is much more immunogenic as compared to IM because skin serves as the essential gate for the entry of pathogens and is always abundant with antigen presenting cells.Various other devices have been designed for efficient DNA delivery like 'Gene Gun'. Along with the gene gun techniques like electroporation and "Particle mediated epidermal delivery" (PMED) are also in use.These are also known as Second generation DNA vaccines. Multiple projects working on DNA vaccine rely on different methods. Various other techniques require the use of naked pDNA wherein it is worked out in a nanoparticle composed of lipids or polymers so that pDNA stability increases and thus resulting in higher cellular uptake.

Attempts to Improve DNA Vaccines

Efficacy of DNA vaccines can be improved by enhancing the immunogenicity of the encoded antigen. This can either be done by adding a genetic adjuvant or by altering the gene encoding the antigen itself. Usually a genetic adjuvant is added to increase the immunogenicity of the antigen and enhance the effects of vaccine. HGMB162 is an example of a genetic adjuvant. Immunogenicity of the antigen can also be increased by codon optimization , addition of signal sequences and genetic fusion to an entire protein referred to as carrier protein.Codon optimization is defined as the technique in which the gene encoding the antigen is edited for optimal transcription and translation in the species that the vaccine is meant for. With the repetition of the genetic code the optimal tRNA for any amino acid differs from species to species. Specifically when innate prokaryotic or viral genes are used in DNA vaccines, codon optimization can considerably enhance its transcription in the eukaryotic cells of the vaccinated host. Addition of signal sequences aims the antigen to various subcellular compartments, hence increasing the immunogenicity of DNA vaccines. Fusion of the antigen to a carrier protein is a device that is often used and required in the construction of DNA vaccines. (Table).

Table: Popular carrier proteins used for fusions with HPV-16 E6 and E7:

Carrier protein	Antigen	Proposed mode of action
Mycobacterium tuberculosis, HSP-70	E7	Provision of CD4+ T-cell help, Increased antigen uptake by DC
Heat shock protein 60	E6, E7	Increased antigen uptake by DC
Calreticulin	E6, E7	Targeting of antigen into the antigen presentation pathway
Extracellular domain of Flt3 ligand	E7	Altered subcellular localisation
HSV VP22	E7	Antigen spreading , Improved antigen stability
TTFC	E6, E7	Increased antigen stability
IP-10	E7	Enhanced antigen presentation, chemoattraction
Invariant chain with PADRE epitope insertion	E6	Provision of CD4+ T cell help
Pseudomonas aeruginosa exotoxin A (domain II)	E7	Enhanced cross presentation
E. coli β-glucoronidase	E7	Enhanced stability

How Safe are DNA Vaccines

Usually DNA vaccines are granted safe for both patient and environment. So far limited local reactivity at the injection site is the only drawback in the studies done. However due to the use of genetic vaccines there is a potential threat that the genome of the plasmid may get integrated into host genome of somatic cells and may lead to the production of transformed cells or oncogenes. To evaluate the potency of the associated risk, the parameter generally set is to compare the integration rate of pDNA with the spontaneous mutation frequency of autologous genes.This mutation rate may vary from person to person but overall, 2 x10 -6 spontaneous gene-inactivating, mutations per gene is usually approved as the standard value.There has been no evidence of elaborated studies carried out for genomic integration of DNA vaccines in humans which might be partly due to hurdles to obtain a biopsy from the administration site. However such studies have been carried out in many lab animals and all these studies have shown that the integration rates are always several folds less than the spontaneous integration rate.

Secondary Role of DNA Vaccines

Although DNA vaccines are still not exposed to human population openly but it has many other roles to carry out. It may be because the plasmid DNA can be used *in vitro* or *in vivo*. This property of DNA vaccines enables them to design monoclonal and polyclonal antibodies. Other experiments that require the use of DNA vaccines are the making of Knock-out mice models and also for developing DNA vaccine libraries to use them to know which genes encode protective antigens without any info about its corresponding protein.

RNA Vaccines

RNA vaccines have been known in the field of vaccine development is quite well known. Over the past 25 years, many clinical trials of DNA in the form of plasmid and viral vector based vaccines have showed us a safe and efficacious way to deliver many foreign antigens. Yet, plenty and satisfactory potency for general efficacy in humans has remained elusive for DNA and RNA vaccines and the practicality of repeated use of viral vectors has been compromised by anti-vector immunity. RNA vaccines, including those based on mRNA and self-amplifying RNA replicons, have the tendency to surpass the restrictions of plasmid DNA and viral vectors. Possible difficulties related to the cost and feasibility of synthesizing RNA vaccines are being addressed, increasing the likelihood that RNA-based vaccines will be commercially available. Proof of concept for RNA vaccines has been shown in humans and the prospects for further progress into commercial products are very motivating.

Transposable Element

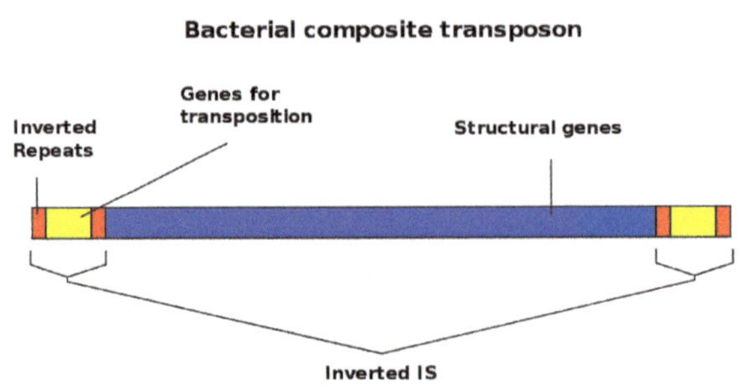

A bacterial DNA transposon

A transposable element (TE or transposon) is a DNA sequence that can change its position within a genome, sometimes creating or reversing mutations and altering the cell's genome size. Transposition often results in duplication of the TE. Barbara McClintock's discovery of these jumping genes earned her a Nobel Prize in 1983.

Transposable elements make up a large fraction of the genome and are responsible for much of the mass of DNA in a eukaryotic cell. It has been shown that TEs are important in genome function and evolution. In *Oxytricha*, which has a unique genetic system, these elements play a critical role in development. Transposons are also very useful to researchers as a means to alter DNA inside a living organism.

There are at least two classes of TEs: Class I TEs or retrotransposons generally function via reverse transcription, while Class II TEs or DNA transposons encode the protein transposase, which they require for insertion and excision, and some of these TEs also encode other proteins.

Discovery

Barbara McClintock discovered the first TEs in maize (*Zea mays*) at the Cold Spring Harbor Laboratory in New York. McClintock was experimenting with maize plants that had broken chromosomes.

In the winter of 1944–1945, McClintock planted corn kernels that were self-pollinated, meaning that the silk (style) of the flower received pollen from its own anther. These kernels came from a long line of plants that had been self-pollinated, causing broken arms on the end of their ninth chromosomes. As the maize plants began to grow, McClintock noted unusual color patterns on the leaves. For example, one leaf had two albino patches of almost identical size, located side by side on the leaf. McClintock hypothesized that during cell division certain cells lost genetic material, while others gained what they had lost. However, when comparing the chromosomes of the current generation of plants with the parent generation, she found certain parts of the chromosome had switched position. This refuted the popular genetic theory of the time that genes were fixed in their position on a chromosome. McClintock found that genes could not only move, but they could also be turned on or off due to certain environmental conditions or during different stages of cell development.

McClintock also showed that gene mutations could be reversed. She presented her report on her findings, and published an article on her discoveries in *Genetics* entitled "Induction of Instability at Selected Loci in Maize."

Her work would be largely dismissed and ignored until the late 1960s–1970s when it would be rediscovered after TEs were found in bacteria. She was awarded a Nobel Prize in Physiology or Medicine in 1983 for her discovery of TEs, more than thirty years after her initial research.

Approximately 90% of the maize genome is made up of TEs, as is 44% of the human genome.

Classification

Transposable elements represent one of several types of mobile genetic elements. TEs are assigned to one of two classes according to their mechanism of transposition, which can be described as either *copy and paste* (Class I TEs) or *cut and paste* (Class II TEs).

Class I (Retrotransposons)

Class I TEs are copied in two stages: first, they are transcribed from DNA to RNA, and the RNA produced is then reverse transcribed to DNA. This copied DNA is then inserted back into the genome at a new position. The reverse transcription step is catalyzed by a reverse transcriptase, which is often encoded by the TE itself. The characteristics of retrotransposons are similar to retroviruses, such as HIV.

Retrotransposons are commonly grouped into three main orders:

- TEs with long terminal repeats (LTRs), which encode reverse transcriptase, similar to retroviruses
- Long interspersed nuclear elements (LINEs, LINE-1s, or L1s), which encode reverse transcriptase but lack LTRs, and are transcribed by RNA polymerase II
- Short interspersed nuclear elements do not encode reverse transcriptase and are transcribed by RNA polymerase III

[Note : Retroviruses can also be considered TEs. For example, after conversion of retroviral RNA into DNA inside a host cell, the newly produced retroviral DNA is integrated into the genome of the host cell. These integrated DNAs are termed *proviruses*. The provirus is a specialized form of eukaryotic retrotransposon, which can produce RNA intermediates that may leave the host cell and infect other cells. The transposition cycle of retroviruses has similarities to that of prokaryotic TEs, suggesting a distant relationship between the two].

Class II (DNA Transposons)

The cut-and-paste transposition mechanism of class II TEs does not involve an RNA intermediate. The transpositions are catalyzed by several transposase enzymes. Some transposases non-specifically bind to any target site in DNA, whereas others bind to specific target sequences. The transposase makes a staggered cut at the target site producing sticky ends, cuts out the DNA transposon and ligates it into the target site. A DNA polymerase fills in the resulting gaps from the sticky ends and DNA ligase closes the sugar-phosphate backbone. This results in target site duplication and the insertion sites of DNA transposons may be identified by short direct repeats (a staggered cut in the target DNA filled by DNA polymerase) followed by inverted repeats (which are important for the TE excision by transposase).

Cut-and-paste TEs may be duplicated if their transposition takes place during S phase of the cell cycle, when a donor site has already been replicated but a target site has not yet been replicated. Such duplications at the target site can result in gene duplication, which plays an important role in genomic evolution.

Not all DNA transposons transpose through the cut-and-paste mechanism. In some cases, a replicative transposition is observed in which a transposon replicates itself to a new target site (e.g. helitron).

Class II TEs comprise less than 2% of the human genome, making the rest Class I.

Autonomous and Non-autonomous

Transposition can be classified as either "autonomous" or "non-autonomous" in both Class I and Class II TEs. Autonomous TEs can move by themselves, whereas non-autonomous TEs require the presence of another TE to move. This is often because dependent TEs lack transposase (for Class II) or reverse transcriptase (for Class I).

Activator element (*Ac*) is an example of an autonomous TE, and dissociation element (*Ds*) is an example of a non-autonomous TE. Without *Ac, Ds* is not able to transpose.

Examples

- The first TEs were discovered in maize (*Zea mays*) by Barbara McClintock in 1948, for which she was later awarded a Nobel Prize. She noticed chromosomal insertions, deletions, and translocations caused by these elements. These changes in the genome could, for example, lead to a change in the color of corn kernels. About 85% of the maize genome consists of TEs. The Ac/Ds system described by McClintock are Class II TEs. Transposition of Ac in tobacco has been demonstrated by B. Baker (Plant Transposable Elements, pp 161–174, 1988, Plenum Publishing Corp., ed. Nelson).
- One family of TEs in the fruit fly *Drosophila melanogaster* are called *P elements*. They seem to have first appeared in the species only in the middle of the twentieth century; within the last 50 years, they spread through every population of the species. Gerald M. Rubin and Allan C. Spradling pioneered technology to use artificial P elements to insert genes into *Drosophila* by injecting the embryo.
- Transposons in bacteria usually carry an additional gene for functions other than transposition, often for antibiotic resistance. In bacteria, transposons can jump

from chromosomal DNA to plasmid DNA and back, allowing for the transfer and permanent addition of genes such as those encoding antibiotic resistance (multi-antibiotic resistant bacterial strains can be generated in this way). Bacterial transposons of this type belong to the Tn family. When the transposable elements lack additional genes, they are known as insertion sequences.

- The most common transposable element in humans is the Alu sequence. It is approximately 300 bases long and can be found between 300,000 and one million times in the human genome. *Alu* alone is estimated to make up 15–17% of the human genome.
- Mariner-like elements are another prominent class of transposons found in multiple species, including humans. The Mariner transposon was first discovered by Jacobson and Hartl in *Drosophila*. This Class II transposable element is known for its uncanny ability to be transmitted horizontally in many species. There are an estimated 14,000 copies of Mariner in the human genome comprising 2.6 million base pairs. The first mariner-element transposons outside of animals were found in *Trichomonas vaginalis*. These characteristics of the Mariner transposon inspired the science fiction novel *The Mariner Project* by Bob Marr.
- Mu phage transposition is the best-known example of replicative transposition.
- Yeast (*Saccharomyces cerevisiae*) genomes contain five distinct retrotransposon families: Ty1, Ty2, Ty3, Ty4 and Ty5.
- A helitron is a TE found in eukaryotes that is thought to replicate by a rolling-circle mechanism.
- Silk fiber is majorly composed of a fibrous protein called fibroin, which has hundreds of tandem repeat motifs derived through transposible elements during evolution, this confers tight packing and high crystallinity that offers tensile strength and lustrous nature to silk. In 2015, *Adarsh Gupta K* et al of Nagaraju's research team at *Centre for DNA Fingerprinting and Diagnostics*, Hyderabad, India discovered the protein structure of Silk Fibroin of Indian golden silkmoth, and published in *Nature* Scientific Reports.

In Disease

TEs are mutagens and their movements are often the causes of genetic disease. They can damage the genome of their host cell in different ways:

- a transposon or a retrotransposon that inserts itself into a functional gene will most likely disable that gene;
- after a DNA transposon leaves a gene, the resulting gap will probably not be repaired correctly;

- multiple copies of the same sequence, such as Alu sequences, can hinder precise chromosomal pairing during mitosis and meiosis, resulting in unequal cross-overs, one of the main reasons for chromosome duplication.

Diseases often caused by TEs include hemophilia A and B, severe combined immunodeficiency, porphyria, predisposition to cancer, and Duchenne muscular dystrophy. *LINE1* (*L1*) TEs that land on the human Factor VIII have been shown to cause haemophilia and insertion of *L1* into the *APC* gene causes colon cancer, confirming that TEs play an important role in disease development.

Additionally, many TEs contain promoters which drive transcription of their own transposase. These promoters can cause aberrant expression of linked genes, causing disease or mutant phenotypes.

Rate of Transposition, Induction and Defense

One study estimated the rate of transposition of a particular retrotransposon, the Ty1 element in *Saccharomyces cerevisiae*. Using several assumptions, the rate of successful transposition event per single Ty1 element came out to be about once every few months to once every few years. Some TEs contain heat-shock like promoters and their rate of transposition increases if the cell is subjected to stress, thus increasing the mutation rate under these conditions, which might be beneficial to the cell.

Cells defend against the proliferation of TEs in a number of ways. These include piRNAs and siRNAs, which silence TEs after they have been transcribed.

If organisms are mostly composed of TEs, one might assume that disease caused by misplaced TEs is very common, but in most cases TEs are silenced through epigenetic mechanisms like DNA methylation, chromatin remodeling and piRNA, such that little to no phenotypic effects nor movements of TEs occur as in some wild-type plant TEs. Certain mutated plants have been found to have defects in methylation-related enzymes (methyl transferase) which cause the transcription of TEs, thus affecting the phenotype.

One hypothesis suggests that only approximately 100 LINE1 related sequences are active, despite their sequences making up 17% of the human genome. In human cells, silencing of LINE1 sequences is triggered by an RNA interference (RNAi) mechanism. Surprisingly, the RNAi sequences are derived from the 5' untranslated region (UTR) of the LINE1, a long terminal which repeats itself. Supposedly, the 5' LINE1 UTR that codes for the sense promoter for LINE1 transcription also encodes the antisense promoter for the miRNA that becomes the substrate for siRNA production. Inhibition of the RNAi silencing mechanism in this region showed an increase in LINE1 transcription.

Evolution

TEs are found in most life forms, and the scientific community is still exploring their evolution and their effect on genome evolution. It is unclear whether TEs originated in the last universal common ancestor, arose independently multiple times, or arose once and then spread to other kingdoms by horizontal gene transfer. While some TEs confer benefits on their hosts, most are regarded as selfish DNA parasites. In this way, they are similar to viruses. Various viruses and TEs also share features in their genome structures and biochemical abilities, leading to speculation that they share a common ancestor.

Because excessive TE activity can damage exons, many organisms have developed mechanisms to inhibit their activity. Bacteria may undergo high rates of gene deletion as part of a mechanism to remove TEs and viruses from their genomes, while eukaryotic organisms typically use RNA interference to inhibit TE activity. Nevertheless, some TEs generate large families often associated with speciation events. Evolution often deactivates DNA transposons, leaving them as introns (inactive gene sequences). In vertebrate animal cells, nearly all 100,000+ DNA transposons per genome have genes that encode inactive transposase polypeptides. In humans, all Tc1-like transposons are inactive. The first synthetic transposon designed for use in vertebrate cells, the Sleeping Beauty transposon system, is a Tc1/mariner-like transposon. It exists in the human genome as an intron and was activated through reconstruction.

Large quantities of TEs within genomes may still present evolutionary advantages, however. Interspersed repeats within genomes are created by transposition events accumulating over evolutionary time. Because interspersed repeats block gene conversion, they protect novel gene sequences from being overwritten by similar gene sequences and thereby facilitate the development of new genes. TEs may also have been co-opted by the vertebrate immune system as a means of producing antibody diversity. The V(D)J recombination system operates by a mechanism similar to that of some TEs.

TEs can contain many types of genes, including those conferring antibiotic resistance and ability to transpose to conjugative plasmids. Some TEs also contain integrons, genetic elements that can capture and express genes from other sources. These contain integrase, which can integrate gene cassettes. There are over 40 antibiotic resistance genes identified on cassettes, as well as virulence genes.

Transposons do not always excise their elements precisely, sometimes removing the adjacent base pairs; this phenomenon is called exon shuffling. Shuffling two unrelated exons can create a novel gene product or, more likely, an intron.

Applications

The first TE was discovered in maize (*Zea mays*) and is named dissociator (*Ds*). Likewise, the first TE to be molecularly isolated was from a plant (snapdragon). Appropriately, TEs have been an especially useful tool in plant molecular biology. Researchers use them as a means of mutagenesis. In this context, a TE jumps into a gene and produces a mutation. The presence of such a TE provides a straightforward means of identifying the mutant allele relative to chemical mutagenesis methods.

Sometimes the insertion of a TE into a gene can disrupt that gene's function in a reversible manner, in a process called insertional mutagenesis; transposase-mediated excision of the DNA transposon restores gene function. This produces plants in which neighboring cells have different genotypes. This feature allows researchers to distinguish between genes that must be present inside of a cell in order to function (cell-autonomous) and genes that produce observable effects in cells other than those where the gene is expressed.

TEs are also a widely used tool for mutagenesis of most experimentally tractable organisms. The Sleeping Beauty transposon system has been used extensively as an insertional tag for identifying cancer genes.

The Tc1/mariner-class of TEs Sleeping Beauty transposon system, awarded Molecule of the Year in 2009, is active in mammalian cells and is being investigated for use in human gene therapy.

TEs are used for the reconstruction of phylogenies by the means of presence/absence analyses.

De Novo Repeat Identification

De novo repeat identification is an initial scan of sequence data that seeks to find the repetitive regions of the genome, and to classify these repeats. Many computer programs exist to perform *de novo* repeat identification, all operating under the same general principles. As short tandem repeats are generally 1–6 base pairs in length and are often consecutive, their identification is relatively simple. Dispersed repetitive elements, on the other hand, are more challenging to identify, due to the fact that they are longer and have often acquired mutations. However, it is important to identify these repeats as they are often found to be transposable elements (TEs).

De novo identification of transposons involves three steps: 1) find all repeats within the genome, 2) build a consensus of each family of sequences, and 3) classify these repeats. There are three groups of algorithms for the first step. One group is referred to as the k-mer approach, where a k-mer is a sequence of length k. In this approach, the genome is scanned for overrepresented k-mers; that is, k-mers that occur more often than is likely based on probability alone. The length k is determined by the type of transposon being searched for. The k-mer approach also allows mismatches, the number of which is determined by the analyst. Some k-mer approach programs use the k-mer as a base, and extend both ends of each repeated k-mer until there is no more similarity between them, indicating the ends of the repeats. Another group of algorithms employs a method called sequence self-comparison. Sequence self-comparison programs use databases such as AB-BLAST to conduct an initial sequence alignment. As these programs find groups of elements that partially overlap, they are useful for finding highly diverged transposons, or transposons with only a small region copied into other parts of the genome. Another group of algorithms follows the periodicity approach. These algorithms perform a Fourier transformation on the sequence data, identifying periodicities, regions that are repeated periodically, and are able to use peaks in the resultant spectrum to find candidate repetitive elements. This method works best for tandem repeats, but can be used for dispersed repeats as well. However, it is a slow process, making it an unlikely choice for genome scale analysis.

The second step of *de novo* repeat identification involves building a consensus of each family of sequences. A consensus sequence is a sequence that is created based on the repeats that comprise a TE family. A base pair in a consensus is the one that occurred most often in the sequences being compared to make the consensus. For example, in a family of 50 repeats where 42 have a T base pair in the same position, the consensus sequence would have a T at this position as well, as the base pair is representative of the family as a whole at that particular position, and is most likely the base pair found in the family's ancestor at that position. Once a consensus sequence has been made for each family, it is then possible to move on to further analysis, such as TE classification and genome masking in order to quantify the overall TE content of the genome.

Adaptive TEs

Transposable elements have been recognized as good candidates for stimulating gene adaptation, through their ability to regulate the expression levels of nearby genes. Combined with their "mobility," transposable elements can be relocated adjacent to their targeted genes, and control the expression levels of the gene, dependent upon the circumstances.

The study conducted in 2008, High Rate of Recent Transposable Element–Induced Adaptation in Drosophila melanogaster, used D. melanogaster that had recently migrated from Africa to other parts of the world, as a basis for studying adaptations caused by transposable elements. Although most of the TEs were located on introns, the experiment showed the significant difference on gene expressions between the population in Africa and other parts of the world. The four TEs that caused the selective sweep were more prevalent in D. melanogaster from temperate climates, leading the researchers to conclude that the selective pressures of the climate prompted genetic adaptation. From this experiment, it has been confirmed that adaptive TEs are prevalent in nature, by enabling organisms to adapt gene expression as a result of new selective pressures.

However, not all effects of adaptive TEs are beneficial to the population. In the research conducted in 2009, A Recent Adaptive Transposable Element Insertion Near Highly Conserved Developmental Loci in Drosophila melanogaster, a TE, inserted between Jheh 2 and Jheh 3, revealed a downgrade in the expression level of both of the genes. Down regulation of such genes has caused drosophila to exhibit extended developmental time and reduced egg to adult viability. Although this adaptation was observed in high frequency in all non-African populations, it was not fixed in any of them. This is not hard to believe, since it is logical for a population to favor higher egg to adult viability, therefore trying to purge the trait caused by this specific TE adaptation.

At the same time, there have been several reports showing the advantageous adaptation caused by TEs. In the research done with silkworms, An Adaptive Transposable Element insertion in the Regulatory Region of the EO Gene in the Domesticated Silkworm, a TE insertion was observed in the cis-regulatory region of the EO gene, which regulates molting hormone 20E, and enhanced expression was recorded. While populations without the TE insert are often unable to effectively regulate hormone 20E under starvation conditions, those with the insert had a more stable development, which resulted in higher developmental uniformity.

These three experiments all demonstrated different ways in which TE insertions can be advantageous or disadvantageous, through means of regulating the expression level of adjacent genes. The field of adaptive TE research is still under development and more findings can be expected in the future.

Lentiviral Vector in Gene Therapy

Lentiviral vectors in gene therapy is a method by which genes can be inserted, modified, or deleted in organisms using lentivirus.

Lentivirus are a family of viruses that are responsible for notable diseases like HIV. The Lentivirus is unique in that it has been the basis of research using viruses in gene therapy. To be effective in gene therapy, there must be insertion, alteration and/or removal of host cell genes. To do this scientists use the Lentivirus' mechanisms of infection to achieve a desired outcome to gene therapy.

Lentivirus

To understand the capabilities of the Lentivirus as a vector, the biology of the infectious process must be understood. The Lentivirus is a retrovirus, meaning it has a single stranded RNA genome with a reverse transcriptase enzyme. Lentiviruses also have a viral envelope with protruding glycoproteins that aid in attachment to the host cell's outer membrane. The virus contains a reverse transcriptase molecule found to perform transcription of the viral genetic material upon entering the cell. Within the viral genome are DNA sequences that code for specific enzymes effecting the incorporation of the viral DNA into the host cell genome. The "gag" domain codes for the structural components of the virus like the capsid, the matrix, nucleoproteins. The "pol" domain codes for the reverse transcriptase and integrase enzymes. Lastly, the "env" domain of the viral genome encodes for the glycoproteins and envelope on the surface of the virus.

There are multiple steps involved in the infection and replication of a Lentivirus in a host cell. In the first step the virus uses its surface glycoproteins for attachment to the outer surface of a cell. More specifically, Lentiviruses attach to the CD40 ligand glycoproteins on the surface of T-lymphocyte cells. The viral material is then injected into the host cell's cytoplasm. Within the cytoplasm the viral reverse transcriptase enzyme performs reverse transcription of the viral RNA genome to create a viral DNA genome. The viral DNA is then sent into the nucleus of the host cell where it is then incorporated into the host cell's genome with the help of the viral enzyme integrase. From there the host cell performs transcription and translation to create viral particles and assemble virions which then burst from the host cell when enough are made.

Use as a Vector

Some of the experimental uses for the Lentivirus vector have been in gene therapy of diseases like Diabetes mellitus, Murine haemophilia A, prostate cancer, chronic granulomatous disease, and vascular gene therapy. Therapy requires manipulation of the Lentivirus genes and structure for delivery of specific genes to alter the course of the disease. Researchers and scientists do this through the use of transinfection, which is the action of infecting a cell with a microbe to gain different genes or desired traits.

Applications

Vascular Transplants

In a study designed to enhance the outcomes of vascular transplant through vascular endothelial cell gene therapy, the third generation of Lentivirus showed to be effective in the delivery of genes to moderate venous grafts and transplants in procedures like coronary artery bypass. Because the virus has been adapted to lose most of its genome, the virus becomes safer and more effective in transplanting the required genes into the host cell. A draw back to this therapy is explained in the study that long-term gene expression may require the use of promoters and can aid in a greater trans-gene expression. The researchers accomplished this by the addition of self-inactivating plasmids and creating a more universal tropism by pseudotyping a vesicular stomatis virus glycoprotein.

Chronic Granulomatous

In chronic granulomatous disease immune functioning is deficient as a result of the loss of nicotinomide adenine dinucleotide phosphate oxidase (NADP) in phagocyte cells, which aids in lipid and nucleic acid synthesis. If this becomes deficient, the bodies immune responses become weakened. Study performed in mice emphasizes the use of lineage-specific lentiviral vectors for the production of NADP. Scientists developed this strain of lentivirus by transinfecting 293T cells with pseudotyped virus with the vesicular stomatitis G protein. The viral vector's responsibility was to increase the gene synthesis and production of NADP in these phagocytic cells. They did this to create an affinity for myeloid cells.

Prostate Cancer

With prostate cancer, the Lentivirus is transformed by being bound to trastuzumab to attach to androgen-sensitive LNCaP and castration-resistant C4-2 human prostate cancer cell lines. These two cells are primarily responsible for secretion of excess human epidermal growth factor receptor 2 (HER-2), which is a hormone linked to prostate cancer. By attaching to these cells and changing their genomes, the Lentivirus can slow down, and even kill, the cancer causing cells. Researchers caused the specificity of the vector by manipulating the Fab region of the viral genome and pseudotyped it with the Sindbis virus.

Haemophilia A

Haemophilia A has also been studied in gene therapy with a lentiviral vector in mice. The vector targets the haematopoietic cells in order to increase the amount of factor VIII, which is affected in haemophilia A. But this continues to be a subject of study as the lentivirus vector was not completely successful in achieving this goal. They did this by trans-infecting the virus in a 293T cell, creating a virus known as 2bF8 expressing generation of viral vectors.

Rheumatoid Arthritis

Studies have also found that injection of a lentiviral vector with IL-10 expressing genes in utero in mice can suppress, and prevent, rheumatoid arthritis and create new cells with constant gene expression. This contributes to the data on stem cells and in utero inoculation of viral vectors for gene therapy. The target for the viral vector in this study, were the synovial cells. Normally functioning synovial cells produce TNFα and IL-1.

Diabetes Mellitus

Like many of the in utero studies, the lentiviral vector gene therapy for diabetes mellitus is more effective in utero as the stem cells that become affected by the gene therapy create new cells with the new gene created by the actual viral intervention. The vector targets the cells within the pancreas to add insulin secreting genes to help control diabetes mellitus. Vectors were cloned using a cytomegalovirus promoter and then cotransinfected in the 293T cell.

Vehicles for Gene Transfer

The very first step in designing a viral vector is to know the sequences needed for the assembly of virus particles, packaging of genome into the particles, and transduction of gene of interest to the targeted cells. The dispensable segment of the genome is deleted in order to use the same space for the gene of interest. Moreover removal of the dispensable part reduces the pathogenicity and immunogenicity of the virus particles. Viral vectors are broadly divided into two types; namely integrating (adeno-associated virus) and non-integrating (adenovirus).

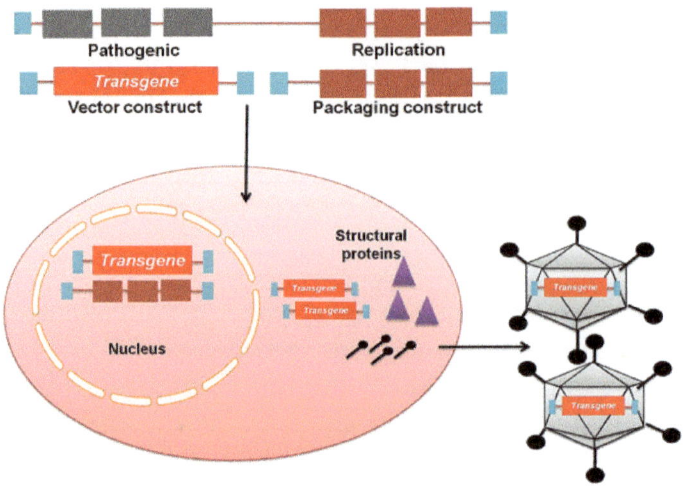

Basic strategy to design a viral vector

Virus genome codes for pathogenic and replication fragments. The pathogenic segment may be replaced by transgene. The cells are transfected with transgene and packaging construct which assembles to form vector. Alternatively, the packaging constructs can also be stably expressed in stable cell line. The stable cell line containing packaging proteins is used to transfect the viral vector containing transgene under the control of tissue specific promoter.

Retrovirus

Retroviruses are a large family of enveloped RNA viruses found in all vertebrates. The enveloped virus particle contains two copies of the viral RNA genome, which are surrounded by a cone-shaped core. The viral RNA contains three essential genes, *gag* , *pol* , and *env* , and is flanked by long terminal repeats (LTRs). The *env* gene encodes for the envelope glycoproteins (gP120 and gP41), which mediates virus entry.

Functions of *gag* , *pol* , and *env*

- Gag protein is proteolytically processed into
- MA (matrix)
- CA (capsid)
- NC (nucleocapsid)
- Pol protein encodes enzymes
- PR (protease)

- RT (Reverse Transcriptase which has both DNA polymerase and RNase H activities)
- IN (Integrase)
- Env protein encodes
- SU surface glycoprotein (gP120)
- TM transmembrane protein (gP41)

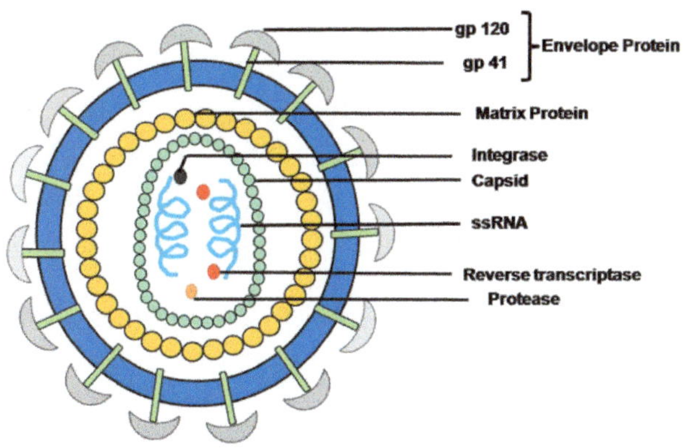

Schematic representation of a retrovirus

The use of each coreceptor corresponds to viruses with different biological properties and pathogenicity. Viruses isolated at the beginning of infection use the CCR5 co-receptor, which is the major coreceptor for macrophage-tropic strains (M-tropic). In full-blown AIDS cases, new viral species appear with high level of replication, cytopathic effects and they use the coreceptor CXCR4, which is the major receptor for T-cell strains (T-cell tropic). There are also dual tropic viruses that can use both CXCR4 and CCR5 coreceptors and alternative chemokine coreceptors. The fusion between the viral membrane and the cellular membrane involves binding of CD4 receptor with gp120 and a change in conformation of gp41, which enables it to insert into the cellular phospholipids bilayer. The retrovirus that binds to the receptor of mice cells are called ECOTROPIC virus while those binding to both human and mice is called as AMPHOTROPIC virus.

Everse transcriptase is an enzyme present in retrovirus (eg HIV) that converts RNA to DNA. The phenomenon is also called as TEMINISM.

Retrovirus

A retrovirus is a single-stranded positive-sense RNA virus with a DNA intermediate and, as an obligate parasite, targets a host cell. Once inside the host cell cytoplasm, the virus uses its own reverse transcriptase enzyme to produce DNA from its RNA genome, the reverse of the usual pattern, thus *retro* (backwards). The new DNA is then incorporated into the host cell genome by an integrase enzyme, at which point the retroviral DNA is referred to as a provirus. The host cell then treats the viral DNA as part of its own genome, translating and transcribing the viral genes along with the cell's own genes, producing the proteins required to assemble new copies of the virus. It is difficult to detect the virus until it has infected the host. At that point, the infection will persist indefinitely.

In most viruses, DNA is transcribed into RNA, and then RNA is translated into protein. However, retroviruses function differently, as their RNA is reverse-transcribed into DNA, which is integrated into the host cell's genome (when it becomes a provirus), and then undergoes the usual transcription and translational processes to express the genes carried by the virus. The information contained in a retroviral gene is thus used to generate the corresponding protein via the sequence: RNA → DNA → RNA → polypeptide. This extends the fundamental process identified by Francis Crick (one gene-one peptide) in which the sequence is DNA → RNA → peptide (proteins are made of one or more polypeptide chains; for example, haemoglobin is a four-chain peptide).

Retroviruses are valuable research tools in molecular biology, and they have been used successfully in gene delivery systems.

Structure

Virions of retroviruses consist of enveloped particles about 100 nm in diameter. The virions also contain two identical single-stranded RNA molecules 7–10 kilobases in length. Although virions of different retroviruses do not have the same morphology or biology, all the virion components are very similar.

The main virion components are:

- Envelope: composed of lipids (obtained from the host plasma membrane during budding process) as well as glycoprotein encoded by the env gene. The retroviral envelope serves three distinct functions: protection from the extracellular environment via the lipid bilayer, enabling the retrovirus to enter/exit host cells through endosomal membrane trafficking, and the ability to directly enter cells by fusing with their membranes.
- RNA: consists of a dimer RNA. It has a cap at the 5' end and a poly(A) tail at the 3' end. The RNA genome also has terminal noncoding regions, which are important in replication, and internal regions that encode virion proteins for gene expression. The 5' end includes four regions, which are R, U5, PBS, and L. The R region is a short repeated sequence at each end of the genome used during the reverse transcription to ensure correct end-to-end transfer in the growing chain. U5, on the other hand, is a short unique sequence between R and PBS. PBS (primer binding site) consists of 18 bases complementary to 3' end of tRNA primer. L region is an untranslated leader region that gives the signal for packaging of the genome RNA. The 3' end includes 3 regions, which are PPT

(polypurine tract), U3, and R. The PPT is a primer for plus-strand DNA synthesis during reverse transcription. U3 is a sequence between PPT and R, which serves as a signal that the provirus can use in transcription. R is the terminal repeated sequence at 3' end.

- Proteins: consisting of gag proteins, protease (PR), pol proteins, and env proteins.
    - Group-specific antigen (gag) proteins are major components of the viral capsid, which are about 2000–4000 copies per virion.
    - Protease is expressed differently in different viruses. It functions in proteolytic cleavages during virion maturation to make mature gag and pol proteins.
    - Pol proteins are responsible for synthesis of viral DNA and integration into host DNA after infection.
    - Env proteins play a role in association and entry of virions into the host cell. Possessing a functional copy of an env gene is what makes retroviruses distinct from retroelements. The ability of the retrovirus to bind to its target host cell using specific cell-surface receptors is given by the surface component (SU) of the Env protein, while the ability of the retrovirus to enter the cell via membrane fusion is imparted by the membrane-anchored trans-membrane component (TM). Thus it is the Env protein that enables the retrovirus to be infectious.

Multiplication

A retrovirus has a membrane containing glycoproteins, which are able to bind to a receptor protein on a host cell. There are two strands of RNA within the cell that have three enzymes: protease, reverse transcriptase, and integrase (1). The first step of replication is the binding of the glycoprotein to the receptor protein (2). Once these have been bound, the cell membrane degrades, becoming part of the host cell, and the RNA strands and enzymes enter the cell (3). Within the cell, reverse transcriptase creates a complementary strand of DNA from the retrovirus RNA and the RNA is degraded; this strand of DNA is known as cDNA (4). The cDNA is then replicated, and the two strands form a weak bond and enter the nucleus (5). Once in the nucleus, the DNA is integrated into the host cell's DNA with the help of integrase (6). This cell can either stay dormant, or RNA may be synthesized from the DNA and used to create the proteins for a new retrovirus (7). Ribosome units are used to transcribe the mRNA of the virus into the amino acid sequences which can be made into proteins in the rough endoplasmic reticulum. This step will also make viral enzymes and capsid proteins (8). Viral RNA will be made in the nucleus. These pieces are then gathered together and are pinched off of the cell membrane as a new retrovirus (9).

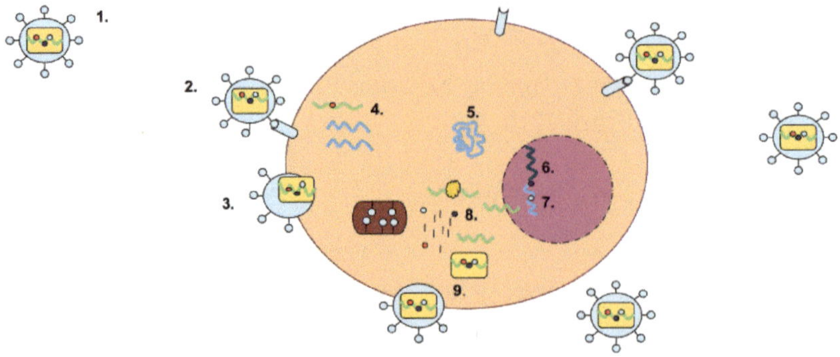

When retroviruses have integrated their own genome into the germ line, their genome is passed on to a following generation. These endogenous retroviruses (ERVs), contrasted with exogenous ones, now make up 5-8% of the human genome. Most insertions have no known function and are often referred to as "junk DNA". However, many endogenous retroviruses play important roles in host biology, such as control of gene transcription, cell fusion during placental development in the course of the germination of an embryo, and resistance to exogenous retroviral infection. Endogenous retroviruses have also received special attention in the research of immunology-related pathologies, such as autoimmune diseases like multiple sclerosis, although endogenous retroviruses have not yet been proven to play any causal role in this class of disease.

While transcription was classically thought to occur only from DNA to RNA, reverse transcriptase transcribes RNA into DNA. The term "retro" in retrovirus refers to this reversal (making DNA from RNA) of the central dogma of molecular biology. Reverse transcriptase activity outside of retroviruses has been found in almost all eukaryotes, enabling the generation and insertion of new copies of retrotransposons into the host genome. These inserts are transcribed by enzymes of the host into new RNA molecules that enter the cytosol. Next, some of these RNA molecules are translated into viral proteins. For example, the *gag* gene is translated into molecules of the capsid protein, the *pol* gene is translated into molecules of reverse transcriptase, and the *env* gene is translated into molecules of the envelope protein. It is important to note that a retrovirus must "bring" its own reverse transcriptase in its capsid, otherwise it is unable to utilize the enzymes of the infected cell to carry out the task, due to the unusual nature of producing DNA from RNA.

Industrial drugs that are designed as protease and reverse transcriptase inhibitors are made such that they target specific sites and sequences within their respective enzymes. However these drugs can quickly become ineffective due to the fact that the gene sequences that code for the protease and the reverse transcriptase quickly mutate. These changes in bases cause specific codons and sites with the enzymes to change and thereby avoid drug targeting by losing the sites that the drug actually targets.

Because reverse transcription lacks the usual proofreading of DNA replication, a retrovirus mutates very often. This enables the virus to grow resistant to antiviral pharmaceuticals quickly, and impedes the development of effective vaccines and inhibitors for the retrovirus.

One difficulty faced with some retroviruses, such as the Moloney retrovirus, involves the requirement for cells to be actively dividing for transduction. As a result, cells such as neurons are very resistant to infection and transduction by retroviruses. This gives rise to a concern that insertional mutagenesis due to integration into the host genome might lead to cancer or leukemia. This is unlike *Lentivirus*, a genus of *Retroviridae*, which are able to integrate their RNA into the genome of non-dividing host cells.

Transmission

- Cell-to-cell
- Fluids
- Airborne, like the Jaagsiekte sheep retrovirus.

Provirus

This DNA can be incorporated into host genome as a provirus that can be passed on to progeny cells. The retrovirus DNA is inserted at random into the host genome. Because of this, it can be inserted into oncogenes. In this way some retroviruses can convert normal cells into cancer cells. Some provirus remains latent in the cell for a long period of time before it is activated by the change in cell environment.

Early Evolution

Studies of retroviruses led to the first demonstrated synthesis of DNA from RNA templates, a fundamental mode for transferring genetic material that occurs in both eukaryotes and prokaryotes. It has been speculated that the RNA to DNA transcription processes used by retroviruses may have first caused DNA to be used as genetic material. In this model, the RNA world hypothesis, cellular organisms adopted the more chemically stable DNA when retroviruses evolved to create DNA from the RNA templates.

Gene Therapy

Gammaretroviral and lentiviral vectors for gene therapy have been developed that mediate stable genetic modification of treated cells by chromosomal integration of the transferred vector genomes. This technology is of use, not only for research purposes, but also for clinical gene therapy aiming at the long-term correction of genetic defects, e.g., in stem and progenitor cells. Retroviral vector particles with tropism for various target cells have been designed. Gammaretroviral and lentiviral vectors have so far been used in more than 300 clinical trials, addressing treatment options for various diseases. Retroviral mutations can be developed to make transgenic mouse models to study various cancers and their metastatic models.

Cancer

Retroviruses that cause tumor growth include *Rous sarcoma virus* and *Mouse mammary tumor virus*. Cancer can be triggered by proto-oncogenes that were mistakenly incorporated into proviral DNA or by the disruption of cellular proto-oncogenes. Rous sarcoma virus contains the src gene that triggers tumor formation. Later it was found that a similar gene in cells is involved in cell signaling, which was most likely excised with the proviral DNA. Nontransforming viruses can randomly insert their DNA into proto-oncogenes, disrupting the expression of proteins that regulate the cell cycle. The promoter of the provirus DNA can also cause over expression of regulatory genes.

Classification

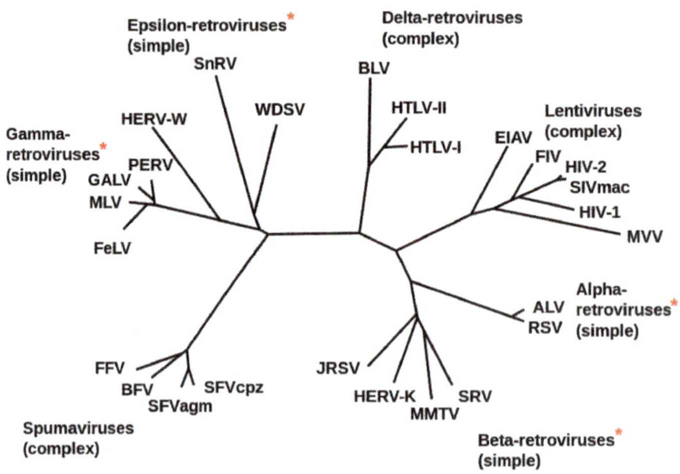

Phylogeny of Retroviruses

Exogenous

These are infectious RNA-containing viruses which are transmitted from human to human.

The following genera are included here:

- Genus *Alpharetrovirus*; type species: *Avian leukosis virus*; others include *Rous sarcoma virus*
- Genus *Betaretrovirus*; type species: *Mouse mammary tumour virus*
- Genus *Gammaretrovirus*; type species: *Murine leukemia virus*; others include *Feline leukemia virus*

- Genus *Deltaretrovirus*; type species: *Bovine leukemia virus*; others include the cancer-causing *Human T-lymphotropic virus*
- Genus *Epsilonretrovirus*; type species: *Walleye dermal sarcoma virus*
- Genus *Lentivirus*; type species: *Human immunodeficiency virus 1*; others include *Simian, Feline* immunodeficiency viruses
- Genus *Spumavirus*; type species: *Simian foamy virus*

These were previously divided into three subfamilies (*Oncovirinae*, *Lentivirinae*, and *Spumavirinae*), but are now divided into two: *Orthoretrovirinae* and *Spumaretrovirinae*. The term oncovirus is now commonly used to describe a cancer-causing virus.

Retroviruses were in 2 groups of the Baltimore classification.

Group VI Viruses

All members of Group VI use virally encoded reverse transcriptase, an RNA-dependent DNA polymerase, to produce DNA from the initial virion RNA genome. This DNA is often integrated into the host genome, as in the case of retroviruses and pseudoviruses, where it is replicated and transcribed by the host.

Group VI includes:

- Family *Metaviridae*
- Family *Pseudoviridae*
- Family *Retroviridae* — Retroviruses, e.g. *HIV*

Group VII Viruses

Both families in Group VII have DNA genomes contained within the invading virus particles. The DNA genome is transcribed into both mRNA, for use as a transcript in protein synthesis, and pre-genomic RNA, for use as the template during genome replication. Virally encoded reverse transcriptase uses the pre-genomic RNA as a template for the creation of genomic DNA.

Group VII includes:

- Family *Hepadnaviridae* — e.g. *Hepatitis B virus*
- Family *Caulimoviridae* — e.g. *Cauliflower mosaic virus*

Endogenous

Endogenous retroviruses are not formally included in this classification system, and are broadly classified into three classes, on the basis of relatedness to exogenous genera:

- Class I are most similar to the gammaretroviruses
- Class II are most similar to the betaretroviruses and alpharetroviruses
- Class III are most similar to the spumaviruses.

Treatment

Antiretroviral drugs are medications for the treatment of infection by retroviruses, primarily HIV. Different classes of antiretroviral drugs act on different stages of the HIV life cycle. Combination of several (typically three or four) antiretroviral drugs is known as highly active anti-retroviral therapy (HAART).

Treatment of Veterinary Retroviruses

Feline leukemia virus and *Feline immunodeficiency virus* infections are treated with biologics, including the only immunomodulator currently licensed for sale in the United States, Lymphocyte T-Cell Immune Modulator (LTCI).

- Virion attachment to a specific cell surface receptor
- Virion penetration into the cell
- Reverse transcription of ther genome.
- Transfer of viral DNA to the infected cell nucleus
- Integration of viral DNA randomly to the cellular DNA to form the provirus
- Viral RNA synthesis by cellular RNA polymerase II using the proviral template
- Transcripts processing to viral genome and mRNAs
- Virus protein synthesis
- Assembly and budding from cell surface
- Processing of capsid proteins

After binding to its receptor (CD4 receptor on T cells), the viral capsid containing the RNA genome enters the cell through membrane fusion. The viral RNA genome is subsequently converted into a double-stranded viral DNA by the viral enzyme reverse transcriptase. The viral DNA is heavily associated with viral proteins like nucleocapsid, reverse transcriptase, and integrase, and translocates to the nucleus where the viral enzyme integrase mediates integration of the viral DNA into the host cell genome to form PROVIRUS.

Adenoviridae

Adenoviruses (members of the family *Adenoviridae*) are medium-sized (90–100 nm), nonenveloped (without an outer lipid bilayer) viruses with an icosahedral nucleocapsid containing a double stranded DNA genome. Their name derives from their initial isolation from human adenoids in 1953.

They have a broad range of vertebrate hosts; in humans, more than 50 distinct adenoviral serotypes have been found to cause a wide range of illnesses, from mild respiratory infections in young children (known as the common cold) to life-threatening multi-organ disease in people with a weakened immune system.

Virology

Group: dsDNA

Order: Unassigned

- Family: Adenoviridae
    - Genus: Atadenovirus
- Bovine atadenovirus D
- Duck atadenovirus A
- *Ovine atadenovirus D*
- Possum atadenovirus A
- Snake atadenovirus A
    - Genus: Aviadenovirus
- Falcon aviadenovirus A
- *Fowl aviadenovirus A*
- Fowl aviadenovirus B
- Fowl aviadenovirus C
- Fowl aviadenovirus D
- Fowl aviadenovirus E
- Goose aviadenovirus A
- Turkey aviadenovirus B

- Genus: Ichtadenovirus
  - *Sturgeon ichtadenovirus A*
- Genus: Mastadenovirus
  - Bat mastadenovirus A
  - Bat mastadenovirus B
  - Bovine mastadenovirus A
  - Bovine mastadenovirus B
  - Bovine mastadenovirus C
  - Canine mastadenovirus A
  - Equine mastadenovirus A
  - Equine mastadenovirus B
  - Human mastadenovirus A
  - Human mastadenovirus B
  - *Human mastadenovirus C*
  - Human mastadenovirus D
  - Human mastadenovirus E
  - Human mastadenovirus F
  - Human mastadenovirus G
  - Murine mastadenovirus A
  - Murine mastadenovirus B
  - Murine mastadenovirus C
  - Ovine mastadenovirus A
  - Ovine mastadenovirus B
  - Porcine mastadenovirus A
  - Porcine mastadenovirus B
  - Porcine mastadenovirus C
  - Simian mastadenovirus A

- Tree shrew mastadenovirus A
    - Genus: Siadenovirus
- *Frog siadenovirus A*
- Great tit siadenovirus A
- Raptor siadenovirus A
- Skua siadenovirus A
- Turkey siadenovirus A

Classification

This family contains the following genera:

- Genus *Atadenovirus*; type species: *Ovine atadenovirus D*
- Genus *Aviadenovirus*; type species: *Fowl aviadenovirus A*
- Genus *Ichtadenovirus*; type species: *Sturgeon ichtadenovirus A*
- Genus *Mastadenovirus* (including all human adenoviruses); type species: *Human mastadenovirus C*
- Genus *Siadenovirus*; type species: *Frog siadenovirus A*

Diversity

Classification of *Adenoviridae* can be complex.

In humans, there are 57 accepted human adenovirus types (HAdV-1 to 57) in seven species (Human adenovirus A to G):

- A: 12, 18, 31
- B: 3, 7, 11, 14, 16, 21, 34, 35, 50, 55
- C: 1, 2, 5, 6, 57
- D: 8, 9, 10, 13, 15, 17, 19, 20, 22, 23, 24, 25, 26, 27, 28, 29, 30, 32, 33, 36, 37, 38, 39, 42, 43, 44, 45, 46, 47, 48, 49, 51, 53, 54, 56
- E: 4
- F: 40, 41
- G: 52

Different types/serotypes are associated with different conditions:

- respiratory disease (mainly species HAdV-B and C)
- conjunctivitis (HAdV-B and D)
- gastroenteritis (HAdV-F types 40, 41, HAdV-G type 52)
- obesity or adipogenesis (HAdV-A type 31, HAdV-C type 5, HAdV-D types 9, 36, 37)

When not restricting the subject to human viruses, *Adenoviridae* can be divided into five genera: *Mastadenovirus*, *Aviadenovirus*, *Atadenovirus*, *Siadenovirus*, and *Ichtadenovirus*.

Structure

Adenoviruses represent the largest nonenveloped viruses. They are able to be transported through the endosome (i.e., envelope fusion is not necessary). The virion also has a unique "spike" or fiber associated with each penton base of the capsid that aids in attachment to the host cell via the receptor on the surface of the host cell.

In 2010, scientists announced that they had solved the structure of the human adenovirus at the atomic level, making the largest high-resolution model ever. The virus is composed of around 1 million amino acid residues and weighs around 150 MDa.

Genome

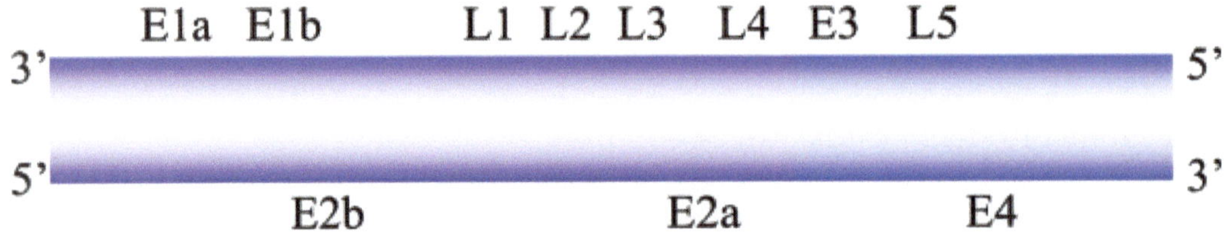

Schematic diagram of the linear adenovirus genome, showing Early genes (E) and Late genes (L).

The adenovirus genome is linear, non-segmented double-stranded (ds) DNA that is between 26 and 48 Kbp. This allows the virus to theoretically carry 22 to 40 genes. Although this is significantly larger than other viruses in its Baltimore group, it is still a very simple virus and is heavily reliant on the host cell for survival and replication. An interesting feature of this viral genome is that it has a terminal 55 kDa protein associated with each of the 5' ends of the linear dsDNA. These are used as primers in viral replication and ensure that the ends of the virus' linear genome are adequately replicated.

Replication

Adenoviruses possess a linear dsDNA genome and are able to replicate in the nucleus of vertebrate cells using the host's replication machinery.

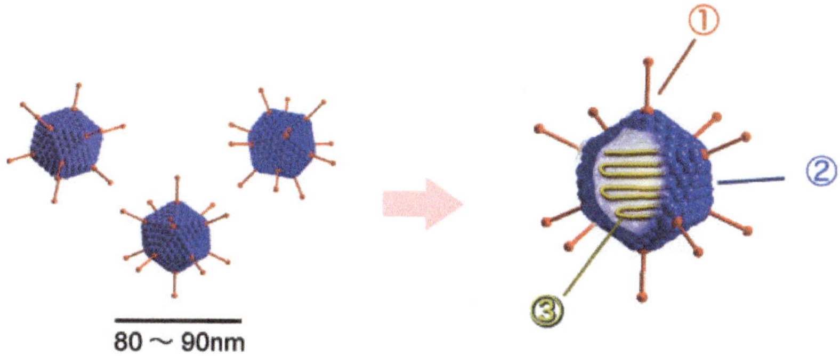

The structure of adenovirus. 1 = penton capsomeres 2 = hexon capsomeres, and 3= viral genome (linear dsDNA)

Entry of adenoviruses into the host cell involves two sets of interactions between the virus and the host cell. Most of the action occurs at the vertices. Entry into the host cell is initiated by the knob domain of the fiber protein binding to the cell receptor. The two currently established receptors are: CD46 for the group B human adenovirus serotypes and the coxsackievirus adenovirus receptor (CAR) for all other serotypes. There are some reports suggesting MHC molecules and sialic acid residues functioning in this capacity as well. This is followed by a secondary interaction, where a motif in the penton base protein interacts with an integrin molecule. It is the co-receptor interaction that stimulates entry of the adenovirus. This co-receptor molecule is αv integrin. Binding to αv integrin results in endocytosis of the virus particle via clathrin-coated pits. Attachment to αv integrin stimulates cell signaling and thus induces actin polymerization resulting in entry of the virion into the host cell within an endosome.

Once the virus has successfully gained entry into the host cell, the endosome acidifies, which alters virus topology by causing capsid components to disband. These changes, as well as the toxic nature of the pentons, destroy the endosome, resulting in the movement of the virion into the cytoplasm. With the help of cellular microtubules, the virus is transported to the nuclear pore complex, whereby the adenovirus particle disassembles. Viral DNA is subsequently released, which can enter the nucleus via the nuclear pore. After this the DNA associates with histone molecules. Thus, viral gene expression can occur and new virus particles can be generated.

The adenovirus life cycle is separated by the DNA replication process into two phases: an early and a late phase. In both phases, a primary transcript that is alternatively spliced to generate monocistronic mRNAs compatible with the host's ribosome is generated, allowing for the products to be translated.

The early genes are responsible for expressing mainly non-structural, regulatory proteins. The goal of these proteins is threefold: to alter the expression of host proteins that are necessary for DNA synthesis; to activate other virus genes (such as the virus-encoded DNA polymerase); and to avoid premature death of the infected cell by the host-immune defenses (blockage of apoptosis, blockage of interferon activity, and blockage of MHC class I translocation and expression).

Some adenoviruses under specialized conditions can transform cells using their early gene products. E1A (binds Retinoblastoma tumor suppressor protein) has been found to immortalize primary cells *in vitro* allowing E1B (binds p53 tumor suppressor) to assist and stably transform the cells. Nevertheless, they are reliant upon each other to successfully transform the host cell and form tumors.

DNA replication separates the early and late phases. Once the early genes have liberated adequate virus proteins, replication machinery, and replication substrates, replication of the adenovirus genome can occur. A terminal protein that is covalently bound to the 5' end of the adenovirus genome acts as a primer for replication. The viral DNA polymerase then uses a strand displacement mechanism, as opposed to the conventional Okazaki fragments used in mammalian DNA replication, to replicate the genome.

The late phase of the adenovirus lifecycle is focused on producing sufficient quantities of structural protein to pack all the genetic material produced by DNA replication. Once the viral components have successfully been replicated, the virus is assembled into its protein shells and released from the cell as a result of virally induced cell lysis.

Multiplicity Reactivation

Adenovirus is capable of multiplicity reactivation (MR) (Yamamoto and Shimojo, 1971). MR is the process by which two, or more, virus genomes containing lethal damage interact within the infected cell to form a viable virus genome. Such MR was demonstrated for adenovirus 12 after virions were irradiated with UV light and allowed to undergo multiple infection of host cells. In a review, numerous examples of MR in different viruses were described, and it was suggested that MR is a common form of sexual interaction that provides the survival advantage of recombinational repair of genome damages.

Epidemiology

Transmission

Adenoviruses are unusually stable to chemical or physical agents and adverse pH conditions, allowing for prolonged survival outside of the body and water. Adenoviruses are spread primarily via respiratory droplets, however they can also be spread by fecal routes. Research into the molecular mechanisms underlying adenoviral transmission provide empirical evidence in support of the hypothesis that cellular receptors for adenovirus and coxsackievirus (CARs) are needed to transport adenoviruses into certain naive/progenitor cell types.

Humans

Humans infected with adenoviruses display a wide range of responses, from no symptoms at all to the severe infections typical of Adenovirus serotype 14.

Animals

Bat adenovirus TJM (Bt-AdV-TJM) is a novel species of the Mastadenovirus genus isolated from Myotis and Scotophilus kuhlii in China. It is a double stranded DNA virus with no RNA sequence. It is most closely related to the tree shrew and canine AdVs.

Two types of canine adenoviruses are well known, type 1 and 2. Type 1 causes infectious canine hepatitis, a potentially fatal disease involving vasculitis and hepatitis. Type 1 infection can also cause respiratory and eye infections. *Canine adenovirus 2* (CAdV-2) is one of the potential causes of kennel cough. Core vaccines for dogs include attenuated live CAdV-2, which produces immunity to CAdV-1 and CAdV-2. CAdV-1 was initially used in a vaccine for dogs, but corneal edema was a common complication.

Adenovirus in Reptiles is poorly understood, but research is currently in progress.

Adenoviruses are also known to cause respiratory infections in horses, cattle, pigs, sheep, and goats. *Equine adenovirus 1* can also cause fatal disease in immunocompromised Arabian foals, involving pneumonia and destruction of pancreatic and salivary gland tissue. *Tupaia adenovirus* (TAV) (Tree shrew adenovirus 1) has been isolated from tree shrews.

Otarine adenovirus 1 has been isolated from sea lions.

The Fowl adenoviruses are associated with many disease conditions in domestic fowl like Inclusion body hepatitis, Hydropericardium syndrome, Egg drop syndrome, Quail bronchitis and many respiratory conditions.They have also been isolated from wild kites (Milvus migrans).

Titi monkey adenovirus was isolated from a colony of monkeys.

Prevention

In the past, US military recruits were vaccinated against two serotypes of adenovirus, with a corresponding decrease in illnesses caused by those serotypes. That vaccine is no longer manufactured. The U.S. Army Medical Research and Materiel Command announced on 31 October 2011 that a new adenovirus vaccine, which replaces the older version that has been out of production for over a decade, was shipped to basic training sites Oct. 18, 2011. More information is available here.

Good hygiene, including handwashing, is still the best way to avoid picking up the adenovirus from an infected person.

Infections

Most infections with adenovirus result in infections of the upper respiratory tract. Adenovirus infections often show up as conjunctivitis, tonsilitis (which may look exactly like strep throat and cannot be distinguished from strep except by throat culture), an ear infection, or croup. Adenoviruses, types 40 and 41 can also cause gastroenteritis. A combination of conjunctivitis and tonsilitis is particularly common with adenovirus infections. Some children (especially small ones) can develop adenovirus bronchiolitis or pneumonia, both of which can be severe. In babies, adenoviruses can also cause coughing fits that look almost exactly like whooping cough. Adenoviruses can also cause viral meningitis or encephalitis. Rarely, adenovirus can cause hemorrhagic cystitis (inflammation of the urinary bladder—a form of urinary tract infection — with blood in the urine).

Most people recover from adenovirus infections by themselves, but people with immunodeficiency sometimes die of adenovirus infections, and — rarely — even previously healthy people can die of these infections.

Adenoviruses are often transmitted by expectorate, but can also be transmitted by contact with an infected person, or by virus particles left on objects such as towels and faucet handles. Some people with adenovirus gastroenteritis may shed the virus in their stools for months after getting over the symptoms. The virus can be passed through water in swimming pools that do not have enough chlorine in them. As with many other illnesses, good handwashing is one way to inhibit the spread of adenoviruses from one person to another. Heat and bleach will kill adenoviruses on objects.

Treatment

There are no antiviral drugs to treat adenoviral infections, so treatment is largely directed at the symptoms (such as acetaminophen for fever). A doctor may give antibiotic eyedrops for conjunctivitis, while awaiting results of bacterial cultures, and to help prevent secondary bacterial infections. Currently, there is no adenovirus vaccine available to the general public, but a vaccine is available for the United States military for Types 4 and 7.

Use in Treatment of Unrelated Diseases

Adenoviruses have long been a popular viral vector for gene therapy due to their ability to affect both replicating and non-replicating cells, accommodate large transgenes, and code for proteins without integrating into the host cell genome. More specifically, they are used as a vehicle to administer targeted therapy, in the form of recombinant DNA or protein. This therapy has been found especially useful in treating cystic fibrosis. Specific modifications on fibre proteins are used to target Adenovirus to certain cell types; a major effort is made to limit hepatotoxicity and prevent multiple organ failure. Adenovirus dodecahedron can qualify as a potent delivery platform for foreign antigens to human myeloid dendritic cells (MDC), and that it is efficiently presented by MDC to M1-specific CD8+ T lymphocytes.

In China, oncolytic adenovirus is an approved cancer treatment.

Viral Replication

Viral replication is the formation of biological viruses during the infection process in the target host cells. Viruses must first get into the cell before viral replication can occur. From the perspective of the virus, the purpose of viral replication is to allow production and survival of its kind. By generating abundant copies of its genome and packaging these copies into viruses, the virus is able to continue infecting new hosts. Replication between viruses is greatly varied and depends on the type of genes involved in them. Most DNA viruses assemble in the nucleus while most RNA viruses develop solely in cytoplasm.

Viral Production / Replication

Viruses multiply only in living cells. The host cell must provide the energy and synthetic machinery and the low molecular-weight precursors for the synthesis of viral proteins and nucleic acids.

The virus replication occurs in seven stages, namely;

1. Adsorption,
2. Entry,
3. Uncoating,
4. Transcription / mRNA production,
5. Synthesis of virus components,
6. Virion assembly and
7. Release(Liberation Stage).

Adsorption

The virus attaches itself to the Cell Membrane of the host cell. It then injects its DNA or RNA into the host to initiate infection.

Entry

The cell membrane of the host cell invaginates the virus particle, enclosing it in a pinocytotic vacoule. This protects the cell from antibodies like in the case of the HIV virus.

Uncoating

Cell enzymes (from lysosomes) strip off the virus protein coat. This releases or renders accessible the virus nucleic acid or genome.

Transcription / mRNA Production

For some RNA viruses, the infecting RNA produces messenger RNA (mRNA). This is translation of the genome into protein produces. For others with negative stranded RNA and DNA, viruses are produced by transcription then translation.

The mRNA is used to instruct the host cell to make virus components. The virus takes advantage of the existing cell structures to replicate itself.

Synthesis of Virus Components

The following components are manufactured by the virus through the host's existing organelles:

- Viral protein synthesis: virus mRNA is translated on cell ribosomes into two types of virus protein.
- Structural: the proteins which make up the virus particle are manufactured and assembled.
- Non – structural: not found in particle, mainly enzymes for virus genome replication.
- Viral nucleic acid synthesis (genome replication) new virus genome are synthesized, templates are either the parental genome or with single stranded nucleic acid genomes, newly formed complementary strands. By a virus called polymerate or replicate in some DNA viruses by a cell enzyme. This is done in rapidly dividing cells.

Virion Assembly

A virion is simply an active or intact virus particle. In this stage, newly synthesized genome(nucleic acid), and proteins are assembled to form new virus particles.

This may take place in the cell's nucleus, cytoplasm, or at plasma membrane for most developed viruses.

Release (Liberation Stage)

The viruses, now being mature are released by either sudden rupture of the cell, or gradual extrusion(budding) of enveloped viruses through the cell membrane.

The new viruses may invade or attack other cells, or remain dormant in the cell.

Baltimore Classification System

Viruses are classed into 7 types of genes, each of which has its own families of viruses, which in turn have differing replication strategies themselves. David Baltimore, a Nobel Prize-winning biologist, devised a system called the Baltimore Classification System to classify different viruses based on their unique replication strategy. There are seven different replication strategies based on this system (Baltimore Class I, II, III, IV, V, VI, VII). The seven classes of viruses are listed here briefly and in generalities.

Class 1: Double Stranded DNA Viruses

This type of virus usually must enter the host nucleus before it is able to replicate. Some of these viruses require host cell polymerases to replicate their genome, while others, such as adenoviruses or herpes viruses, encode their own replication factors. However, in either cases, replication of the viral genome is highly dependent on a cellular state permissive to DNA replication and, thus, on the cell cycle. The virus may induce the cell to forcefully undergo cell division, which may lead to transformation of the cell and, ultimately, cancer. An example of a family within this classification is the Adenoviridae

There is only one well-studied example in which a class 1 family of viruses does not replicate within the nucleus. This is the Poxvirus family, which comprises highly pathogenic viruses that infect vertebrates.

Class 2: Single-stranded DNA Viruses

Viruses that fall under this category include ones that are not as well-studied, but still do pertain highly to vertebrates. Two examples include the Circoviridae and Parvoviridae. They replicate within the nucleus, and form a double-stranded DNA intermediate during replication. A human Circovirus called TTV is included within this classification and is found in almost all humans, infecting them asymptomatically in nearly every major organ.

Class 3: Double-stranded RNA Viruses

Like most viruses with RNA genomes, double-stranded RNA viruses do not rely on host polymerases for replication to the extent that viruses with DNA genomes do. Double-stranded RNA viruses are not as well-studied as other classes. This class includes two major families, the Reoviridae and Birnaviridae. Replication is monocistronic and includes individual, segmented genomes, meaning that each of the genes codes for only one protein, unlike other viruses, which exhibit more complex translation.

Classes 4 & 5: Single-stranded RNA Viruses

These viruses consist of two types, however both share the fact that replication is primarily in the cytoplasm, and that replication is not as dependent on the cell cycle as that of DNA viruses. This class of viruses is also one of the most-studied types of viruses, alongside the double-stranded DNA viruses.

Class 4: Single-stranded RNA Viruses - Positive-sense

The positive-sense RNA viruses and indeed all genes defined as positive-sense can be directly accessed by host ribosomes to immediately form proteins. These can be divided into two groups, both of which replicate in the cytoplasm:

- Viruses with polycistronic mRNA where the genome RNA forms the mRNA and is translated into a polyprotein product that is subsequently cleaved to form the mature proteins. This means that the gene can utilize a few methods in which to produce proteins from the same strand of RNA, all in the sake of reducing the size of its gene.

- Viruses with complex transcription, for which subgenomic mRNAs, ribosomal frameshifting and proteolytic processing of polyproteins may be used. All of which are different mechanisms with which to produce proteins from the same strand of RNA.

Examples of this class include the families Coronaviridae, Flaviviridae, and Picornaviridae.

Class 5: Single-stranded RNA viruses - Negative-sense

The negative-sense RNA viruses and indeed all genes defined as negative-sense cannot be directly accessed by host ribosomes to immediately form proteins. Instead, they must be transcribed by viral polymerases into the "readable" complementary positive-sense. These can also be divided into two groups:

- Viruses containing nonsegmented genomes for which the first step in replication is transcription from the negative-stranded genome by the viral RNA-dependent RNA polymerase to yield monocistronic mRNAs that code for the various viral proteins. A positive-sense genome copy that serves as template for production of the negative-strand genome is then produced. Replication is within the cytoplasm.

- Viruses with segmented genomes for which replication occurs in the cytoplasm and for which the viral RNA-dependent RNA polymerase produces monocistronic mRNAs from each genome segment.

Examples in this class include the families Orthomyxoviridae, Paramyxoviridae, Bunyaviridae, Filoviridae, and Rhabdoviridae (which includes rabies).

Class 6: Positive-sense Single-stranded RNA Viruses that Replicate Through a DNA Intermediate

A well-studied family of this class of viruses include the retroviruses. One defining feature is the use of reverse transcriptase to convert the positive-sense RNA into DNA. Instead of using the RNA for templates of proteins, they use DNA to create the templates, which is spliced into the host genome using integrase. Replication can then commence with the help of the host cell's polymerases

Class 7: Double-stranded DNA viruses that Replicate Through a single-stranded RNA Intermediate

This small group of viruses, exemplified by the Hepatitis B virus, have a double-stranded, gapped genome that is subsequently filled in to form a covalently closed circle (ccc DNA) that serves as a template for production of viral mRNAs and a subgenomic RNA. The pregenome RNA serves as template for the viral reverse transcriptase and for production of the DNA genome.

Retroviral Vectors

Among the viral vectors used for gene therapy trials, retrovirus is the most commonly used RNA virus. Retroviruses are the first to develop for the application of gene therapy related studies. Retroviruses are the RNA containing enveloped viruses and are classified into oncoretroviruses, lentiviruses, and spumaviruses.

Retroviruses are enveloped viruses having two copies of ssRNA genome. The viral genome codes for three essential proteins; namely *gag* , *pol* , and *env*. The genome is flanked with long terminal repeats (LTR) in its terminal end. The *gag* gene in retroviruses codes for capsid, matrix, and nucleocapsid proteins. The *pol* gene codes for viral enzymes protease, integrase, and reverse transcriptase. The *env* gene codes for surface glycoproteins, which mediates the virus entry into the cells. Oncoretroviruses only codes for these three proteins while lentiviruses and spumaviruses are more complex in nature (discussed in later chapter).

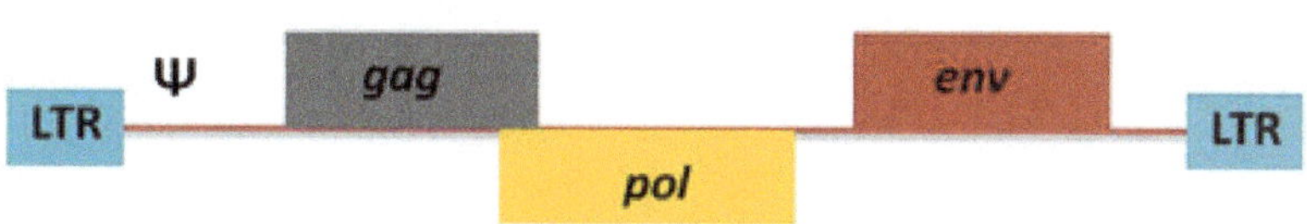

Schematic diagram of retrovirus

Entry of virus particle is accomplished by the fusion of viral surface glycoprotein with the host cell receptor. The viral RNA genome is converted into DNA by the enzyme reverse transcriptase. The DNA after combining with the viral proteins then migrates to nucleus and integrates with the host cell chromosome with the help of enzyme integrase. The transcription of the viral mRNA starts from LTR using host cell transcription factors. The new virus particles are formed after assembling of two copies of ssRNA along with the essential enzyme inside core. The mature virus particles are released by budding from the host cell membrane.

Retroviral vectors are based on replication deficient retroviruses. The vectors are mainly derived from Rous sarcoma virus, avian leukosis virus, and murine leukemia virus. The retroviral vectors are made by replacing the viral proteins with the gene of interest driven by a tissue specific promoter. The vector also contains the long terminal repeat (LTR) which is an essential packaging signal. In addition, the vector also contains essential enzyme such as reverse transcriptase and integrase. Vector RNA production is achieved by either the LTR or promoters present upstream to the transgene. The packaging of the vector is completed by the incorporation of viral structural proteins in *trans* from packaging cell lines. Alternatively, the vector can also be produced by the transfection of plasmid expressing the structural proteins. The latter method is less time consuming as it avoids the use of packaging cell line. Viruses are recovered from the supernatants of actively growing producer cells.

Major issue of using retroviral vectors is the possibility of the defective genome to be recombined with the host cell chromosome. The concern leads to the development of self inactivating retroviral vectors by using tissue specific promoter for transgene transcription. In self inactivating vectors, the transcription of the transgene is carried out by using an internal promoter instead of LTR. Apart for gene delivery, retroviral vectors are extensively used for many other applications. Retroviruses have ability to integrate with the host cell chromosome; the property is explored to express the protein of interest in a cell. Long term expression of a protein in a suitable cell is a very useful way to make a stable cell line. The replication depends on actively dividing cells making these an important tool to manipulate stem cells and tumor cells. On the other hand lentiviral vectors are used for gene delivery in many organs including brain, eye, liver, muscles, and hematopoietic cells.

The transduction of retroviral vector is limited because of its ability to have limited cellular tropism. The cellular tropism of the retrovirus is broadened by the incorporation of envelope from related or unrelated viruses, making them pseudotype virus. Incorporation of vesicular stomatitis virus glycoproteins into the retrovirus virion or the envelope of murine leukemia virus allows the broad host range to the transducing vectors. Pseudotyping of retrovirus by the use of lyssavirus glycoprotein makes it transducible to the brain while incorporation of surface protein of ebola virus helps in transducing airway epithelium.

Adeno-associated Virus

Adeno-associated virus (AAV) is a small virus which infects humans and some other primate species. AAV is not currently known to cause disease. The virus causes a very mild immune response, lending further support to its apparent lack of pathogenicity. Gene therapy vectors using AAV can infect both dividing and quiescent cells and persist in an extrachromosomal state without integrating into the genome of the host cell, although in the native virus some integration of virally carried genes into the host genome does occur. These features make AAV a very attractive candidate for creating viral vectors for gene therapy, and for the creation of isogenic human disease models. Recent human clinical trials using AAV for gene therapy in the retina have shown promise.

AAV belongs to the genus *Dependoparvovirus*, which in turn belongs to the family *Parvoviridae*. The virus is a small (20 nm) replication-defective, nonenveloped virus.

History

The adeno-associated virus (AAV), previously thought to be a contaminant in adenovirus preparations, was first identified as a dependovirus in the 1960s in the laboratories of Bob Atchison at Pittsburgh and Wallace Rowe at NIH. Serological studies in humans subsequently indicated that, despite being present in people infected by helper viruses such as adenovirus or herpes virus, AAV itself did not cause any disease.

Use in Gene Therapy

Advantages and Drawbacks

Wild-type AAV has attracted considerable interest from gene therapy researchers due to a number of features. Chief amongst these is the virus's apparent lack of pathogenicity. It can also infect non-dividing cells and has the ability to stably integrate into the host cell genome at a specific site (designated AAVS1) in the human chromosome 19. The feature makes it somewhat more predictable than retroviruses, which present the threat of a random insertion and of mutagenesis, which is sometimes followed by development of a cancer. The AAV genome integrates most frequently into the site mentioned, while random incorporations into the genome take place with a negligible frequency. Development of AAVs as gene therapy vectors, however, has eliminated this integrative capacity by removal of the *rep* and *cap* from the DNA of the vector. The desired gene together with a promoter to drive transcription of the gene is inserted between the inverted terminal repeats (ITR) that aid in concatemer formation in the nucleus after the single-stranded vector DNA is converted by host cell DNA polymerase complexes into double-stranded DNA. AAV-based gene therapy vectors form episomal concatemers in the host cell nucleus. In non-dividing cells, these concatemers remain intact for the life of the host cell. In dividing cells, AAV DNA is lost through cell division, since the episomal DNA is not replicated along with the host cell DNA. Random integration of AAV DNA into the host genome is detectable but occurs at very low frequency. AAVs also present very low immunogenicity, seemingly restricted to generation of neutralizing antibodies, while they induce no clearly defined cytotoxic response.

This feature, along with the ability to infect quiescent cells present their dominance over adenoviruses as vectors for human gene therapy.

Use of the virus does present some disadvantages. The cloning capacity of the vector is relatively limited and most therapeutic genes require the complete replacement of the virus's 4.8 kilobase genome. Large genes are, therefore, not suitable for use in a standard AAV vector. Options are currently being explored to overcome the limited coding capacity. The AAV ITRs of two genomes can anneal to form head to tail concatamers, almost doubling the capacity of the vector. Insertion of splice sites allows for the removal of the ITRs from the transcript.

Because of AAV's specialized gene therapy advantages, researchers have created an altered version of AAV termed self-complementary adeno-associated virus (scAAV). Whereas AAV packages a single strand of DNA and must wait for its second strand to be synthesized, scAAV packages two shorter strands that are complementary to each other. By avoiding second-strand synthesis, scAAV can express more quickly, although as a caveat, scAAV can only encode half of the already limited capacity of AAV. Recent reports suggest that scAAV vectors are more immunogenic than single stranded adenovirus vectors, inducing a stronger activation of cytotoxic T lymphocytes.

The humoral immunity instigated by infection with the wild type is thought to be a very common event. The associated neutralising activity limits the usefulness of the most commonly used serotype AAV2 in certain applications. Accordingly, the majority of clinical trials currently under way involve delivery of AAV2 into the brain, a relatively immunologically privileged organ. In the brain, AAV2 is strongly neuron-specific.

Clinical Trials

To date, AAV vectors have been used in over 117 clinical trials worldwide, approximately 5.6% of virus-vectored gene-therapy trials. Recently, promising results have been obtained from Phase 1 and Phase 2 trials for a number of diseases, including Leber's congenital amaurosis, Hemophilia, congestive heart failure, spinal muscular atrophy, lipoprotein lipase deficiency, and Parkinson's disease.

Selected Clinical Trials using AAV-Based Vectors					
Indication	Gene	Route of administration	Phase	Subject number	Status
Cystic fibrosis	*CFTR*	Lung, via aerosol	I	12	Complete
	CFTR	Lung, via aerosol	II	38	Complete
	CFTR	Lung, via aerosol	II	100	Complete
Hemophilia B	*FIX*	Intramuscular	I	9	Complete
	FIX	Hepatic artery	I	6	Ended
Arthritis	*TNFR:Fc*	Intraarticular	I	1	Ongoing

Hereditary emphysema	*AAT*	Intramuscular	I	12	Ongoing
Leber's congenital amaurosis	*RPE65*	Subretinal	I–II	Multiple	Several ongoing and complete
Age-related macular degeneration	*sFlt-1*	Subretinal	I–II	24	Ongoing
Duchenne muscular dystrophy	*SGCA*	Intramuscular	I	10	Ongoing
Parkinson's disease	*GAD65, GAD67*	Intracranial	I	12	Complete
Canavan disease	*AAC*	Intracranial	I	21	Ongoing
Batten disease	*CLN2*	Intracranial	I	10	Ongoing
Alzheimer's disease	*NGF*	Intracranial	I	6	Ongoing
Spinal muscular atrophy	*SMN1*	Intravenous	I–II	15	Ongoing
Congestive heart failure	*SERCA2a*	Intra-coronary	IIb	250	Ongoing

AAV structure

AAV Genome, Transcriptome and Proteome

The AAV genome is built of single-stranded deoxyribonucleic acid (ssDNA), either positive- or negative-sensed, which is about 4.7 kilobase long. The genome comprises inverted terminal repeats (ITRs) at both ends of the DNA strand, and two open reading frames (ORFs): *rep* and *cap*. The former is composed of four overlapping genes encoding Rep proteins required for the AAV life cycle, and the latter contains overlapping nucleotide sequences of capsid proteins: VP1, VP2 and VP3, which interact together to form a capsid of an icosahedral symmetry.

ITR Sequences

The Inverted Terminal Repeat (ITR) sequences comprise 145 bases each. They were named so because of their symmetry, which was shown to be required for efficient multiplication of the AAV genome. The feature of these sequences that gives them this property is their ability to form a hairpin, which contributes to so-called self-priming that allows primase-independent synthesis of the second DNA strand. The ITRs were also shown to be required for both integration of the AAV DNA into the host cell genome (19th chromosome in humans) and rescue from it, as well as for efficient encapsidation of the AAV DNA combined with generation of a fully assembled, deoxyribonuclease-resistant AAV particles.

With regard to gene therapy, ITRs seem to be the only sequences required *in cis* next to the therapeutic gene: structural (*cap*) and packaging (*rep*) proteins can be delivered *in trans*. With this assumption many methods were established for efficient production of recombinant AAV (rAAV) vectors containing a reporter or therapeutic gene. However, it was also published that the ITRs are not the only elements required *in cis* for the effective replication and encapsidation. A few research groups have identified a sequence designated *cis-acting Rep-dependent element* (CARE) inside the coding sequence of the *rep* gene. CARE was shown to augment the replication and encapsidation when present *in cis*.

Rep Genes and Rep Proteins

On the "left side" of the genome there are two promoters called p5 and p19, from which two overlapping messenger ribonucleic acids (mRNAs) of different length can be produced. Each of these contains an intron which can be either spliced out or not. Given these possibilities, four various mRNAs, and consequently four various Rep proteins with overlapping sequence can be synthesized. Their names depict their sizes in kilodaltons (kDa): Rep78, Rep68, Rep52 and Rep40. Rep78 and 68 can specifically bind the hairpin formed by the ITR in the self-priming act and cleave at a specific region, designated terminal resolution site, within the hairpin. They were also shown to be necessary for the AAVS1-specific integration of the AAV genome. All four Rep proteins were shown to bind ATP and to possess helicase activity. It was also shown that they upregulate the transcription from the p40 promoter (mentioned below), but downregulate both p5 and p19 promoters.

Cap Genes and VP Proteins

The right side of a positive-sensed AAV genome encodes overlapping sequences of three capsid proteins, VP1, VP2 and VP3, which start from one promoter, designated p40. The molecular weights of these proteins are 87, 72 and 62 kiloDaltons, respectively. All three of them are translated from one mRNA. After this mRNA is synthesized, it can be spliced in two different manners: either a longer or shorter intron can be excised resulting in the formation of two pools of mRNAs: a 2.3 kb- and a 2.6 kb-long mRNA pool. Usually, especially in the presence of adenovirus, the longer intron is preferred, so the 2.3-kb-long mRNA represents the so-called "major splice". In this form the first AUG codon, from which the synthesis of VP1 protein starts, is cut out, resulting in a reduced overall level of VP1 protein synthesis. The first AUG codon that remains in the major splice is the initiation codon for VP3 protein. However, upstream of that codon in the same open reading frame lies an ACG sequence (encoding threonine) which is surrounded by an optimal Kozak context. This contributes to a low level of synthesis of VP2 protein, which is actually VP3 protein with additional N terminal residues, as is VP1.

Since the bigger intron is preferred to be spliced out, and since in the major splice the ACG codon is a much weaker translation initiation signal, the ratio at which the AAV structural proteins are synthesized *in vivo* is about 1:1:20, which is the same as in the mature virus particle. The unique fragment at the N terminus of VP1 protein was shown to possess the phospholipase A2 (PLA2) activity, which is probably required for the releasing of AAV particles from late endosomes. Muralidhar *et al.* reported that VP2 and VP3 are crucial for correct virion assembly. More recently, however, Warrington *et al.* showed VP2 to be unnecessary for the complete virus particle formation and an efficient infectivity, and also presented that VP2 can tolerate large insertions in its N terminus, while VP1 can not, probably because of the PLA2 domain presence.

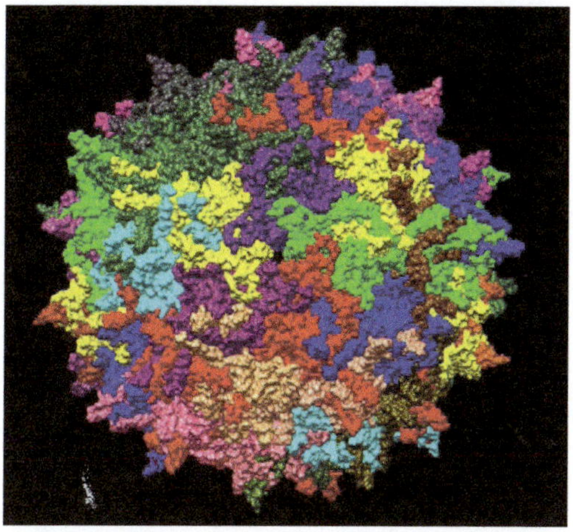

AAV2 capsid, shown as a ribbon diagram, with the back half hidden for clarity. One fivefold symmetry axis is shown center.

The AAV capsid is composed of 60 capsid protein subunits, VP1, VP2, and VP3, that are arranged in an icosahedral symmetry in a ratio of 1:1:10, with an estimated size of 3.9 MegaDaltons. The crystal structure of the VP3 protein was determined by Xie, Bue, *et al.*

AAV Serotypes, Receptors and Native Tropism

Until the 1990s, virtually all AAV biology was studied using AAV serotype 2. However, AAV is highly prevalent in humans and other primates and several serotypes have been isolated from various tissue samples. Serotypes 2, 3, 5, and 6 were discovered in human cells, AAV serotypes 1, 4, and 7–11 in nonhuman primate samples. As of 2006 there have been 11 AAV serotypes described, the 11th in 2004. AAV capsid proteins contain 12 hypervariable surface regions, with most variability occurring in the threefold proximal peaks, but the parvovirus genome in general presents highly conserved replication and structural genes across serotypes. All of the known serotypes can infect cells from multiple diverse tissue types. Tissue specificity is determined by the capsid serotype and pseudotyping of AAV vectors to alter their tropism range will likely be important to their use in therapy.

Serotype 2

Serotype 2 (AAV2) has been the most extensively examined so far. AAV2 presents natural tropism towards skeletal muscles, neurons, vascular smooth muscle cells and hepatocytes.

Three cell receptors have been described for AAV2: heparan sulfate proteoglycan (HSPG), $a_V\beta_5$ integrin and fibroblast growth factor receptor 1 (FGFR-1). The first functions as a primary receptor, while the latter two have a co-receptor activity and enable AAV to enter the cell by receptor-mediated endocytosis. These study results have been disputed by Qiu, Handa, *et al.* HSPG functions as the primary receptor, though its abundance in the extracellular matrix can scavenge AAV particles and impair the infection efficiency.

Serotype 2 and Cancer

Studies have shown that serotype 2 of the virus (AAV-2) apparently kills cancer cells without harming healthy ones. "Our results suggest that adeno-associated virus type 2, which infects the majority of the population but has no known ill effects, kills multiple types of cancer cells yet has no effect on healthy cells," said Craig Meyers, a professor of immunology and microbiology at the Penn State College of Medicine in Pennsylvania in 2005. This could lead to a new anti-cancer agent.

Other Serotypes

Although AAV2 is the most popular serotype in various AAV-based research, it has been shown that other serotypes can be more effective as gene delivery vectors. For instance AAV6 appears much better in infecting airway epithelial cells, AAV7 presents very high transduction rate of murine skeletal muscle cells (similarly to AAV1 and AAV5), AAV8 is superb in transducing hepatocytes and AAV1 and 5 were shown to be very efficient in gene delivery to vascular endothelial cells. In the brain, most AAV serotypes show neuronal tropism, while AAV5 also transduces astrocytes. AAV6, a hybrid of AAV1 and AAV2, also shows lower immunogenicity than AAV2.

Serotypes can differ with the respect to the receptors they are bound to. For example, AAV4 and AAV5 transduction can be inhibited by soluble sialic acids (of different form for each of these serotypes), and AAV5 was shown to enter cells via the platelet-derived growth factor receptor.

AAV Immunology

AAV is of particular interest to gene therapists due to its apparent limited capacity to induce immune responses in humans, a factor which should positively influence vector transduction efficiency while reducing the risk of any immune-associated pathology.

AAV is not considered to have any known role in disease.

Innate

The innate immune response to the AAV vectors has been characterised in animal models. Intravenous administration in mice causes transient production of pro-inflammatory cytokines and some infiltration of neutrophils and other leukocytes into the liver, which seems to sequester a large percentage of the injected viral particles. Both soluble factor levels and cell infiltration appear to return to baseline within six hours. By contrast, more aggressive viruses produce innate responses lasting 24 hours or longer.

Humoral

The virus is known to instigate robust humoral immunity in animal models and in the human population, where up to 80% of individuals are thought to be seropositive for AAV2. Antibodies are known to be neutralising, and for gene therapy applications these do impact on vector transduction efficiency via some routes of administration. As well as persistent AAV specific antibody levels, it appears from both prime-boost studies in animals and from clinical trials that the B-cell memory is also strong. In seropositive humans, circulating IgG antibodies for AAV2 appear to be primarily composed of the IgG1 and IgG2 subclasses, with little or no IgG3 or IgG4 present.

Cell-mediated

The cell-mediated response to the virus and to vectors is poorly characterised, and has been largely ignored in the literature as recently as 2005. Clinical trials using an AAV2-based vector to treat haemophilia B seem to indicate that targeted destruction of transduced cells may be occurring. Combined with data that shows that CD8+ T-cells can recognise elements of the AAV capsid *in vitro*, it appears that there may be a cytotoxic T lymphocyte response to AAV vectors. Cytotoxic responses would imply the involvement of CD4+ T helper cells in the response to AAV and *in vitro* data from human studies suggests that the virus may indeed induce such responses, including both Th1 and Th2 memory responses. A number of candidate T cell stimulating epitopes have been identified within the AAV capsid protein VP1, which may be attractive targets for modification of the capsid if the virus is to be used as a vector for gene therapy.

AAV Infection Cycle

There are several steps in the AAV infection cycle, from infecting a cell to producing new infectious particles:

1. attachment to the cell membrane
2. receptor-mediated endocytosis
3. endosomal trafficking
4. escape from the late endosome or lysosome

5. translocation to the nucleus
6. uncoating
7. formation of double-stranded DNA replicative form of the AAV genome
8. expression of *rep* genes
9. genome replication
10. expression of *cap* genes, synthesis of progeny ssDNA particles
11. assembly of complete virions, and
12. release from the infected cell.

Some of these steps may look different in various types of cells, which, in part, contributes to the defined and quite limited native tropism of AAV. Replication of the virus can also vary in one cell type, depending on the cell's current cell cycle phase.

The characteristic feature of the adeno-associated virus is a deficiency in replication and thus its inability to multiply in unaffected cells. The first factor that was described as providing successful generation of new AAV particles, was the adenovirus, from which the AAV name originated. It was then shown that AAV replication can be facilitated by selected proteins derived from the adenovirus genome, by other viruses such as HSV, or by genotoxic agents, such as UV irradiation or hydroxyurea.

The minimal set of the adenoviral genes required for efficient generation of progeny AAV particles, was discovered by Matsushita, Ellinger *et al.* This discovery allowed for new production methods of recombinant AAV, which do not require adenoviral co-infection of the AAV-producing cells. In the absence of helper virus or genotoxic factors, AAV DNA can either integrate into the host genome or persist in episomal form. In the former case integration is mediated by Rep78 and Rep68 proteins and requires the presence of ITRs flanking the region being integrated. In mice, the AAV genome has been observed persisting for long periods of time in quiescent tissues, such as skeletal muscles, in episomal form (a circular head-to-tail conformation).

Adeno-associated Virus as a Gene Therapy Vector

Different AAV serotypes are reported to use different cell surface receptors for their entry. The specificity of the host cell receptor makes the serotype specific for a particular host. The heparan sulfate proteoglycans present in a variety of cell types is a receptor used by AAV-2. In addition, fibroblast growth factor receptor-1 and integrin $\alpha v \beta 5$ are also reported as a co-receptor for the AAV infection in the host cells. Sialic acid is used as receptor by AAV-4 and AAV-5. Usually after binding to cell receptors, virus particles are internalized by receptor mediated endocytosis and the ssDNA is transported to the host cell nucleus.

The ITRs of AAV contains the self complementary sequences which are hair pin like structure and provides a binding site for the viral *rep* protein. In addition to the *rep* protein mediated virus genome replication, host cell machinery also helps in its replication. The genome replication takes place in the nucleus and both +ve and –ve sense genome is packaged with equal frequency in the virion capsid. The ssDNA is converted to dsDNA form in the nucleus to facilitate the expression of viral proteins and genome replication. The dsDNA is either formed by the complementary DNA strand from another virus or mediated by the host cell machinery. The AAV genome is known to integrate specifically to the chromosome 19 at the location q13.4 by nonhomologous recombination. The integration is mediated by the *rep* proteins, and remains in the cell throughout its life span. Further AAV genome replication depends on the infection of helper virus or their gene products. Upon infection of adeno or herpesvirus, the viral protein regulates the replication and transcription of AAV genome. The new virus particle gets assembled and packaged in the nucleus and released from the cells along with the helper viruses (adeno or herpesvirus).

The first AAV gene therapy vector was generated by replacing internal *rep* and *cap* gene with the transgene. The recombinant virus was produced by the co-infection of helper virus and transfection of plasmid expressing *rep* and *cap* proteins in *trans* . The method still generates the vector having ITR sequences (around 400bps) of the viral genome. The presence of ITR sequences may lead to production of wild type virus in the subsequent generation.

The next generation of AAV contains only part of ITR sequences essential for the production and integration. Essentially the AAV vector system contains a promoter and a transgene flanked by the ITRs, the method prevents the production of replication competent AAV production. The adenovirus E1A, E1B, E2A, VA, and E4 proteins are essential for the production of a successful virus particle. To avoid the use of replication adenovirus particles, scientists have developed a system that requires only E2A, VA, and E4 proteins to produce viable AAV vector. The system requires transfection of E2A, VA, and E4 gene expressing plasmid in 293 cells.

AAV vectors are made by replacing rep/cap genes by transgene and a tissue specific promoter. A packaging construct containing endogenous promoters p5, p40 and p19 are used which provides the regulation for the expression of *rep* and *cap* genes. Helper adenovirus expresses E2, E4 and VA genes in producer cells. The 293 cells are generally used as a producer cell line as it provides the E1 proteins for the generation of AAV particles. Producer cells thus produce non replicating AAV vector. ssDNA is packaged by the structural protein followed by its release after cell lysis.

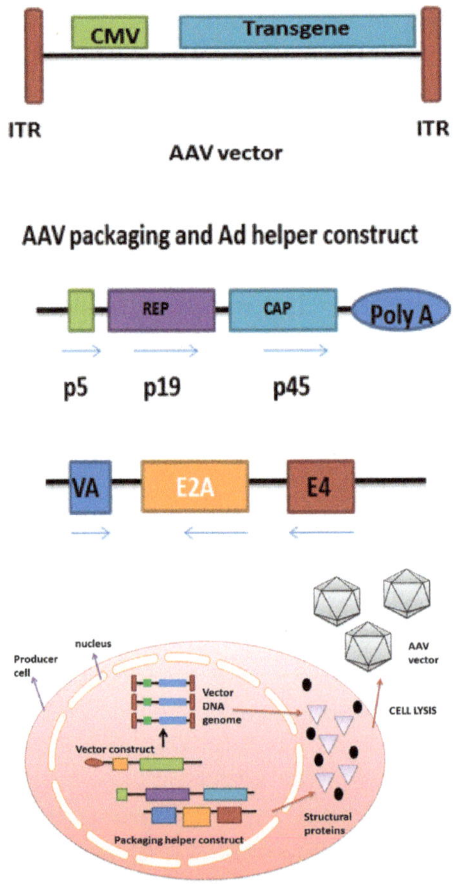

Adeno-associated virus vector development

> Interesting points:
>
> 93 cells are known to express adenovirus E1A and E1B gene products.

Table: Difference between adeno and adeno-associated virus:

Character	Adenovirus	Adeno-associated virus
Integration in host	No	Yes
Expression of vector	Transient	Stable
Transfection efficiency	High	Low
Immune response	High	No
Genome	36kb ds DNA	4.7kb ssDNA
Envelope	No	No
Receptor	CAR	Heparan Sulphate
Cell line	293 HEK	293 HEK
Entry	Caveolae mediated	Clathrin mediated
Proteins	Five early and five late	Four for replicase and three for capsid
Capsid size	40 nm	25 nm

Tropism of adeno-associated Virus

AAV-2 is the most common serotype used to develop a gene therapy vector. The cap protein of AAV-1, -2 and -3 share a great homology with each other and hence are known to use the heparan sulfate receptor. The capsid proteins of AAV-4 and -5 are different from other serotypes and probably involved in using different cellular receptors to infect the susceptible host. The AAV-1 is more tropic towards muscle and liver while AAV-5 is more tropic to retina (eye). AAV-3 is more tropic towards the hematopoietic stem cells. The tropism of AAV serotypes can by modulated by shuffling the capsid proteins. To be more efficient AAV genome should be of dsDNA instead of parental ssDNA, the expression of transgene increases to a large extent after the production of dsDNA from ssDNA by host cell machinery. The dsDNA form of the AAV vector can persist for a long time in a transduced cell by head to tails recombination with ITR sequences.

The major hurdle in AAV mediated gene therapy is regarding the limitation of transgene capacity. Another hurdle is regarding the presence of neutralizing antibodies against the AAV serotypes in the human population because of natural AAV infection. In addition a single injection of AAV can mount a huge humoral immune response that inhibits the second injection of the same serotype. Using AAV serotypes of different capsid protein may overcome this hurdle. The AAV offers a very promising tool for the gene therapy vector and its efficacy and potential is increasing day by day. Currently many gene therapy trials are going on using AAV as a vector expressing different transgene.

References

- Eddy, B. E.; Borman, G. S.; Grubbs, G. E.; Young, R. D. (May 1962). "Identification of the oncogenic substance in rhesus monkey kidney cell culture as simian virus 40". Virology. 17: 65–75. PMID 13889129. doi:10.1016/0042-6822(62)90082-x
- Madigan M, Martinko J (editors) (2006). Brock Biolog of Microorganisms (11th ed.). Prentice Hall. ISBN 0-13-144329-1
- Shah, K; Nathanson, N (January 1976). "Human exposure to SV40: Review and comment". American Journal of Epidemiology. 103 (1): 1–12. PMID 174424
- McClintock, Barbara (June 1950). "The origin and behavior of mutable loci in maize". Proc Natl Acad Sci U S A. 36 (6): 344–55. Bibcode:1950PNAS...36..344M. PMC 1063197 . PMID 15430309. doi:10.1073/pnas.36.6.344
- Sowd, GA; Fanning, E (2012). "A wolf in sheep's clothing: SV40 co-opts host genome maintenance proteins to replicate viral DNA". PLoS Pathogens. 8 (11): e1002994. PMC 3493471 . PMID 23144614. doi:10.1371/journal.ppat.1002994
- Baggott, Mary Tim, ed. (2007). Nature, the Physician, and the Family: Selected Writings of Herbert Ratner, M.D. AuthorHouse. pp. xv–xvii. ISBN 978-1-4184-7510-9
- Eibl, R. H.; Kleihues, P; Jat, P. S.; Wiestler, O. D. (1994). "A model for primitive neuroectodermal tumors in transgenic neural transplants harboring the SV40 large T antigen". The American Journal of Pathology. 144 (3): 556–64. PMC 1887088 . PMID 8129041

- Pershouse, M. A.; Heivly, S; Girtsman, T (2006). "The role of SV40 in malignant mesothelioma and other human malignancies". Inhalation toxicology. 18 (12): 995–1000. PMID 16920674. doi:10.1080/08958370600835377
- Poulin, D. L.; Decaprio, J. A. (2006). "Is There a Role for SV40 in Human Cancer?". Journal of Clinical Oncology. 24 (26): 4356–65. PMID 16963733. doi:10.1200/JCO.2005.03.7101
- McGrayne, Sharon Bertsch (1998). Nobel Prize Women in Science: Their Lives, Struggles, and Momentous Discoveries (2nd ed.). Carol Publishing. p. 165. ISBN 978-0-9702256-0-3
- Shah, K. V. (2007). "SV40 and human cancer: A review of recent data". International Journal of Cancer. Journal International Du Cancer. 120 (2): 215–23. PMID 17131333. doi:10.1002/ijc.22425
- Michod, Richard E.; Bernstein, Harris; Nedelcu, Aurora M. (2008). "Adaptive value of sex in microbial pathogens". Infection, Genetics and Evolution. 8 (3): 267–85. PMID 18295550. doi:10.1016/j.meegid.2008.01.002
- Lohe AR, Moriyama EN, Lidholm DA, Hartl DL (January 1995). "Horizontal transmission, vertical inactivation, and stochastic loss of mariner-like transposable elements". Mol. Biol. Evol. 12 (1): 62–72. PMID 7877497. doi:10.1093/oxfordjournals.molbev.a040191
- Fedoroff, Nina; Botstein, David, eds. (1 January 1992). The Dynamic Genome: Barbara McClintock's Ideas in the Century of Genetics. Cold Spring Harbor Laboratory Press. p. 2. ISBN 978-0-87969-422-7
- Jan Ole Kriegs, Gennady Churakov, Martin Kiefmann, Ursula Jordan, Jürgen Brosius, Jürgen Schmitz, David Penny: Retroposed Elements as Archives for the Evolutionary History of Placental Mammals. In: PLoS Biology. 4, 2006, S. e91, doi:10.1371/journal.pbio.0040091
- Barbanti-Brodano, G; Sabbioni, S; Martini, F; Negrini, M; Corallini, A; Tognon, M (2004). "Simian virus 40 infection in humans and association with human diseases: results and hypotheses". Virology. 318 (1): 1–9. PMID 15015494. doi:10.1016/j.virol.2003.09.004
- González, J., Lenkov, K., Lipatov, M., Macpherson, J. M., & Petrov, D. A. (2008). "High Rate of Recent Transposable Element–Induced Adaptation in Drosophila melanogaster.". PLoS Biology. 6 (10): e251. doi:10.1371/journal.pbio.0060251.

5

Transgenic Animals: A Research Model

The organisms that undergo processes for the modification of their genes are known as transgenic or genetically modified organisms. Mice are the most commonly used genetically modified animals. Easy manipulation of embryos, short gestation periods and availability of complete mouse genome are some of the reasons that mice are used in genetic research. The aspects elucidated in this section are of vital importance, and provide a better understanding of genetically modified organism.

Genetically Modified Organism

A genetically modified organism (GMO) is any organism whose genetic material has been altered using genetic engineering techniques (i.e., a genetically *engineered* organism). GMOs are used to produce many medications and genetically modified foods and are widely used in scientific research and the production of other goods. The term GMO is very close to the technical legal term, 'living modified organism', defined in the Cartagena Protocol on Biosafety, which regulates international trade in living GMOs (specifically, "any living organism that possesses a novel combination of genetic material obtained through the use of modern biotechnology").

GloFish, the first genetically modified animal to be sold as a pet

A more specifically defined type of GMO is a "transgenic organism." This is an organism whose genetic makeup has been altered by the addition of genetic material from an unrelated organism. This should not be confused with the more general way in which "GMO" is used to classify genetically altered organisms, as typically GMOs are organisms whose genetic makeup has been altered without the addition of genetic material from an unrelated organism.

The first genetically modified mouse was created in 1974, and the first plant was produced in 1983.

Production

Genetic modification involves the mutation, insertion, or deletion of genes. Inserted genes usually come from a different species in a form of horizontal gene-transfer. In nature this can occur when exogenous DNA penetrates the cell membrane for any reason. This can be accomplished artificially by:

- attaching the genes to a virus.
- physically inserting the extra DNA into the nucleus of the intended host with a very small syringe.
- using electroporation (that is, introducing DNA from one organism into the cell of another by use of an electric pulse).
- firing small particles from a gene gun.

Other methods exploit natural forms of gene transfer, such as the ability of *Agrobacterium* to transfer genetic material to plants, or the ability of lentiviruses to transfer genes to animal cells.

History

Herbert Boyer (pictured) and Stanley Cohen created the first genetically modified organism in 1973

Humans have domesticated plants and animals since around 12,000 BCE, using selective breeding or artificial selection (as contrasted with natural selection). The process of selective breeding, in which organisms with desired traits (and thus with the desired genes) are used to breed the next generation and organisms lacking the trait are not bred, is a precursor to the modern concept of genetic modification. Various advancements in genetics allowed humans to directly alter the DNA and therefore genes of organisms. In 1972 Paul Berg created the first recombinant DNA molecule when he combined DNA from a monkey virus with that of the lambda virus.

Herbert Boyer and Stanley Cohen made the first genetically modified organism (GMO) in 1973. They took a gene from a bacterium that provided resistance to the antibiotic kanamycin, inserted it into a plasmid and then induced another bacteria to uptake the plasmid. The bacteria was then able to survive in the presence of kanamycin. Boyer and Cohen expressed other genes in bacteria. This included genes from the toad Xenopus laevis in 1974, creating the first GMO expressing a gene from an organism from different kingdom.

In 1974 Rudolf Jaenisch created the first GM animal.

In 1974 Rudolf Jaenisch created a transgenic mouse by introducing foreign DNA into its embryo, making it the world's first transgenic animal. However it took another eight years before transgenic mice were developed that passed the transgene to their offspring. Genetically modified mice were created in 1984 that carried cloned oncogenes, predisposed them to developing cancer. Mice with genes knocked out (knockout mouse) were created in 1989. The first transgenic livestock were produced in 1985 and the first animal to synthesise transgenic proteins in their milk were mice, engineered to produce human tissue plasminogen activator in 1987.

In 1983 the first genetically engineered plant was developed by Michael W. Bevan, Richard B. Flavell and Mary-Dell Chilton. They infected tobacco with *Agrobacterium* transformed with an antibiotic resistance gene and through tissue culture techniques were able to grow a new plant containing the resistance gene. The gene gun was invented in 1987, allowing transformation of plants not susceptible to *Agrobacterium* infection. In 2000, Vitamin A-enriched golden rice, was the first plant developed with increased nutrient value.

In 1976 Genentech, the first genetic engineering company was founded by Herbert Boyer and Robert Swanson; a year later, the company produced a human protein (somatostatin) in *E.coli*. Genentech announced the production of genetically engineered human insulin in 1978. The insulin produced by bacteria, branded humulin, was approved for release by the Food and Drug Administration in 1982. In 1988 the first human antibodies were produced in plants. In 1987, the ice-minus strain of *P. syringae* became the first genetically modified organism to be released into the environment when a strawberry field and a potato field in California were sprayed with it.

The first genetically modified crop, an antibiotic-resistant tobacco plant, was produced in 1982. China was the first country to commercialize transgenic plants, introducing a virus-resistant tobacco in 1992. In 1994 Calgene attained approval to commercially release the Flavr Savr tomato, the first genetically modified food. Also in 1994, the European Union approved tobacco engineered to be resistant to the herbicide bromoxynil, making it the first genetically engineered crop commercialized in Europe. An insect resistant Potato was approved for release in the USA in 1995, and by 1996 approval had been granted to commercially grow 8 transgenic crops and one flower crop (carnation) in 6 countries plus the EU.

In 2010, scientists at the J. Craig Venter Institute, announced that they had created the first synthetic bacterial genome. They named it Synthia and it was the world's first synthetic life form.

The first genetically modified animal to be commercialised was the GloFish, a Zebra fish with a fluorescent gene added that allows it to glow in the dark under ultraviolet light. The first genetically modified animal to be approved for food use was AquAdvantage salmon in 2015. The salmon were transformed with a growth hormone-regulating gene from a Pacific Chinook salmon and a promoter from an ocean pout enabling it to grow year-round instead of only during spring and summer.

Uses

GMOs are used in biological and medical research, production of pharmaceutical drugs, experimental medicine (e.g. gene therapy and vaccines against the Ebola virus), and agriculture (e.g. golden rice, resistance to herbicides), with developing uses in conservation. The term "genetically modified organism" does not always imply, but can include, targeted insertions of genes from one species into another. For example, a gene from a jellyfish, encoding a fluorescent protein called GFP, or green fluorescent protein, can be physically linked and thus co-expressed with mammalian genes to identify the location of the protein encoded by the GFP-tagged gene in the mammalian cell. Such methods are useful tools for biologists in many areas of research, including those who study the mechanisms of human and other diseases or fundamental biological processes in eukaryotic or prokaryotic cells.

Microbes

Bacteria were the first organisms to be modified in the laboratory, due to the relative ease of modifying their genetics.

They continue to be important model organisms for experiments in genetic engineering. In the field of synthetic biology, they have been used to test various synthetic approaches, from synthesizing genomes to creating novel nucleotides.

These organisms are now used for several purposes, and are particularly important in producing large amounts of pure human proteins for use in medicine.

Genetically modified bacteria are used to produce the protein insulin to treat diabetes. Similar bacteria have been used to produce biofuels, clotting factors to treat haemophilia, and human growth hormone to treat various forms of dwarfism.

In addition, various genetically engineered micro-organisms are routinely used as sources of enzymes for the manufacture of a variety of processed foods. These include alpha-amylase from bacteria, which converts starch to simple sugars, chymosin from bacteria or fungi, which clots milk protein for cheese making, and pectinesterase from fungi, which improves fruit juice clarity.

Plants

Transgenic Plants

Kenyans examining insect-resistant transgenic Bt corn

Transgenic plants have been engineered for scientific research, to create new colours in plants, and to create different crops.

In research, plants are engineered to help discover the functions of certain genes. One way to do this is to knock out the gene of interest and see what phenotype develops. Another strategy is to attach the gene to a strong promoter and see what happens when it is over expressed. A common technique used to find out where the gene is expressed is to attach it to GUS or a similar reporter gene that allows visualisation of the location.'

Suntory "blue" rose

After thirteen years of collaborative research, an Australian company – Florigene, and a Japanese company – Suntory, created a blue rose (actually lavender or mauve) in 2004. The genetic engineering involved three alterations – adding two genes, and interfering with another. One of the added genes was for the blue plant pigment delphinidin cloned from the pansy. The researchers then used RNA interference (RNAi) technology to depress all color production by endogenous genes by blocking a crucial protein in color production, called dihydroflavonol 4-reductase) (DFR), and adding a variant of that protein that would not be blocked by the RNAi but that would allow the delphinidin to work. The roses are sold in Japan, the United States, and Canada. Florigene has also created and sells lavender-colored carnations that are genetically engineered in a similar way.

Simple plants and plant cells have been genetically engineered for production of biopharmaceuticals in bioreactors as opposed to cultivating plants in open fields. Work has been done with duckweed *Lemna minor*, the algae *Chlamydomonas reinhardtii* and the moss *Physcomitrella patens*. An Israeli company, Protalix, has developed a method to produce therapeutics in cultured transgenic carrot and tobacco cells. Protalix and its partner, Pfizer, received FDA approval to market its drug Elelyso, a treatment for Gaucher's disease, in 2012.

Genetically Modified Crops

Genetically modified crops (GM crops, or biotech crops) are plants used in agriculture, the DNA of which has been modified using genetic engineering techniques. In most cases the aim is to introduce a new trait to the plant which does not occur naturally in the species. Examples in food crops include resistance to certain pests, diseases, or environmental conditions, reduction of spoilage, or resistance to chemical treatments (e.g. resistance to a herbicide), or improving the nutrient profile of the crop. Examples in non-food crops include production of pharmaceutical agents, biofuels, and other industrially useful goods, as well as for bioremediation.

Farmers have widely adopted GM technology. Between 1996 and 2013, the total surface area of land cultivated with GM crops increased by a factor of 100, from 17,000 square kilometers (4,200,000 acres) to 1,750,000 km² (432 million acres). 10% of the world's croplands were planted with GM crops in 2010. In the US, by 2014, 94% of the planted area of soybeans, 96% of cotton and 93% of corn were genetically modified varieties. In recent years GM crops expanded rapidly in developing countries. In 2013 approximately 18 million farmers grew 54% of worldwide GM crops in developing countries.

Cisgenic Plants

Cisgenesis, sometimes also called intragenesis, is a product designation for a category of genetically engineered plants. A variety of classification schemes have been proposed that order genetically modified organisms based on the nature of introduced genotypical changes rather than the process of genetic engineering.

While some genetically modified plants are developed by the introduction of a gene originating from distant, sexually incompatible species into the host genome, cisgenic plants contain genes that have been isolated either directly from the host species or from sexually compatible species. The new genes are introduced using recombinant DNA methods and gene transfer. Some scientists hope that the approval process of cisgenic plants might be simpler than that of proper transgenics, but it remains to be seen.

Conservation in Plants

Genetically modified organisms have been proposed to aid conservation of plant species threatened by extinction. Many trees face the threat of invasive plants and diseases, such as the emerald ash borer in North American and the fungal disease, Ceratocystis platani, in European plane trees. A suggested solution to increase the resilience of threatened tree species is to genetically modify individuals by transferring resistant genes. Papaya trees are an example of a species that was successfully conserved using genetic modification. The papaya ringspot virus (PRSV) devastated papaya trees in Hawaii in the twentieth century until transgenic papaya plants were given pathogen-derived resistance.

However, genetic modification for conservation in plants remains mainly speculative and further experimentation is needed before the technique can be widely implemented. A main concern with using genetic modification for conservation purposes is that a transgenic species may no longer bear enough resemblance to the original species to truly claim that the original species is being conserved. Instead, the transgenic species may be genetically different enough to be considered a new species, thus diminishing the conservation worth of genetic modification.

Mammals

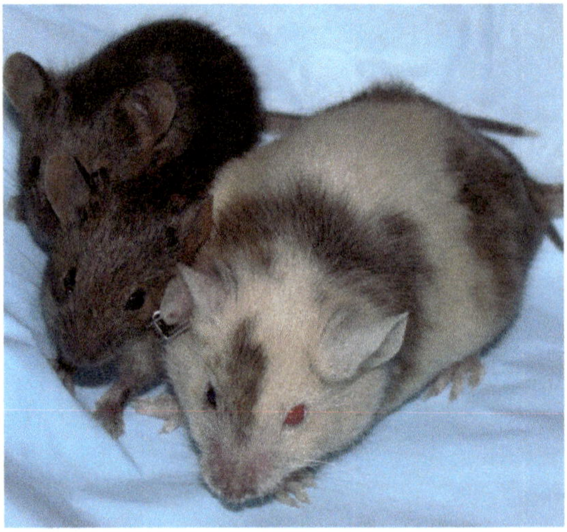

Some chimeras, like the blotched mouse shown, are created through genetic modification techniques like gene targeting.

Genetically modified mammals are an important category of genetically modified organisms. Ralph L. Brinster and Richard Palmiter developed the techniques responsible for transgenic mice, rats, rabbits, sheep, and pigs in the early 1980s, and established many of the first transgenic models of human disease, including the first carcinoma caused by a transgene. The process of genetically engineering animals is a slow, tedious, and expensive process. However, new technologies are making genetic modifications easier and more precise.

The first transgenic (genetically modified) animal was produced by injecting DNA into mouse embryos then implanting the embryos in female mice.

Genetically modified animals currently being developed can be placed into six different broad classes based on the intended purpose of the genetic modification:

1. to research human diseases (for example, to develop animal models for these diseases);

2. to produce industrial or consumer products (fibres for multiple uses);

3. to produce products intended for human therapeutic use (pharmaceutical products or tissue for implantation);

4. to enrich or enhance the animals' interactions with humans (hypo-allergenic pets);

5. to enhance production or food quality traits (faster growing fish, pigs that digest food more efficiently);

6. to improve animal health (disease resistance)

Research use

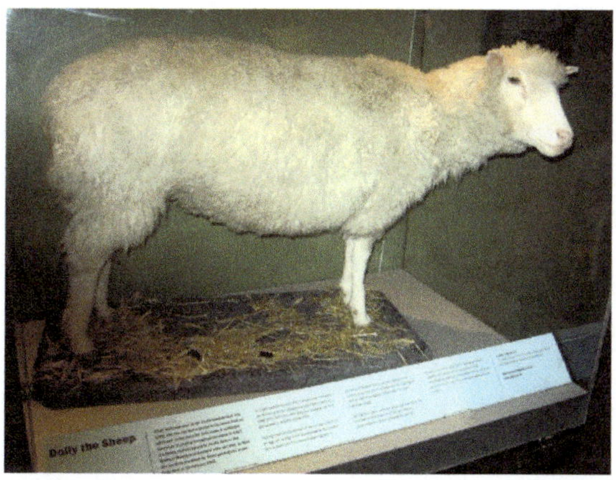

Dolly was a female domestic sheep and the first animal to be cloned from an adult somatic cell

Transgenic animals are used as experimental models to perform phenotypic and for testing in biomedical research.

Genetically modified (genetically engineered) animals are becoming more vital to the discovery and development of cures and treatments for many serious diseases. By altering the DNA or transferring DNA to an animal, we can develop certain proteins that may be used in medical treatment. Stable expressions of human proteins have been developed in many animals, including sheep, pigs, and rats. Human-alpha-1-antitrypsin, which has been tested in sheep and is used in treating humans with this deficiency and transgenic pigs with human-histo-compatibility have been studied in the hopes that the organs will be suitable for transplant with less chances of rejection.

Scientists have genetically engineered several organisms, including some mammals, to include green fluorescent protein (GFP), first observed in the jellyfish, *Aequorea victoria* in 1962, for medical research purposes (Chalfie, Shimoura, and Tsien were awarded the Nobel prize in Chemistry in 2008 for the discovery and development of GFP). For example, fluorescent pigs have been bred to study human organ transplants (xenotransplantation), regenerating ocular photoreceptor cells, and other topics. In 2011 a Japanese-American team created green-fluorescent cats to find therapies for HIV/AIDS and other diseases as feline immunodeficiency virus (FIV) is related to HIV.

In 2009, scientists in Japan announced that they had successfully transferred a gene into a primate species (marmosets) and produced a stable line of breeding transgenic primates for the first time. Their first research target for these marmosets was Parkinson's disease, but they were also considering amyotrophic lateral sclerosis and Huntington's disease.

Producing Human Therapeutics

Herman the Bull, Naturalis, for the production of lactoferrin enhanced milk.

Transgenic pig for cheese production.

Within the field known as pharming, intensive research has been conducted to develop transgenic animals that produce biotherapeutics. On 6 February 2009, the U.S. Food and Drug Administration approved the first human biological drug produced from such an animal, a goat. The drug, ATryn, is an anticoagulant which reduces the probability of blood clots during surgery or childbirth. It is extracted from the goat's milk.

Production or Food Quality Traits

In 2006, a pig was engineered to produce omega-3 fatty acids through the expression of a roundworm gene.

Enviropig was a genetically enhanced line of Yorkshire pigs in Canada created with the capability of digesting plant phosphorus more efficiently than conventional Yorkshire pigs. The project ended in 2012. These pigs produced the enzyme phytase, which breaks down the indigestible phosphorus, in their saliva. The enzyme was introduced into the pig chromosome by pronuclear microinjection. With this enzyme, the animal is able to digest cereal grain phosphorus. The use of these pigs would reduce the potential of water pollution since they excrete from 30 to 70.7% less phosphorus in manure depending upon the age and diet. The lower concentrations of phosphorus in surface runoff reduces algal growth, because phosphorus is the limiting nutrient for algae. Because algae consume large amounts of oxygen, it can result in dead zones for fish.

In 2011, Chinese scientists generated dairy cows genetically engineered with genes from human beings to produce milk that would be the same as human breast milk. This could potentially benefit mothers who cannot produce breast milk but want their children to have breast milk rather than formula. Aside from milk production, the researchers claim these transgenic cows to be identical to regular cows. Two months later scientists from Argentina presented Rosita, a transgenic cow incorporating two human genes, to produce milk with similar properties as human breast milk. In 2012, researchers from New Zealand also developed a genetically engineered cow that produced allergy-free milk.

Goats have been genetically engineered to produce milk with strong spiderweb-like silk proteins in their milk.

Human Gene Therapy

Gene therapy, uses genetically modified viruses to deliver genes that can cure disease in humans. Although gene therapy is still relatively new, it has had some successes. It has been used to treat genetic disorders such as severe combined immunodeficiency, and Leber's congenital amaurosis. Treatments are also being developed for a range of other currently incurable diseases, such as cystic fibrosis, sickle cell anemia, Parkinson's disease, cancer, diabetes, heart disease and muscular dystrophy.

Conservation use

Genetically modified organisms have been used to conserve European wild rabbits in the Iberian peninsula and Australia. In both cases, the genetically modified organism used was a myxoma virus, but for opposite purposes: to protect the endangered population in Europe with immunizations and to regulate the overabundant population in Australia with contraceptives.

In the Iberian peninsula, the European wild rabbit population has experienced a sharp decline from viral diseases and overhunting. To protect the species from viral diseases, the myxoma virus was genetically modified to immunize the rabbits. The European wild rabbit population in Australia faces the opposite problem: lack of natural predators has made the introduced species invasive. The same myxoma virus was genetically modified to lower fertility in the Australian rabbit population.

Fish

Genetically modified fish are used for scientific research and as pets, and are being considered for use as food and as aquatic pollution sensors.

GM fish are widely used in basic research in genetics and development. Two species of fish, zebrafish and medaka, are most commonly modified because they have optically clear chorions (membranes in the egg), rapidly develop, and the 1-cell embryo is easy to see and microinject with transgenic DNA.

The GloFish is a patented brand of genetically modified (GM) fluorescent zebrafish with bright red, green, and orange fluorescent color. Although not originally developed for the ornamental fish trade, it became the first genetically modified animal to become publicly available as a pet when it was introduced for sale in 2003. They were quickly banned for sale in California.

GM fish have been developed with promoters driving an over-production of "all fish" growth hormone for use in the aquaculture industry to increase the speed of development and potentially reduce fishing pressure on wild stocks. This has resulted in dramatic growth enhancement in several species, including salmon, trout and tilapia. AquaBounty Technologies, a biotechnology company working on bringing a GM salmon to market, claims that their GM AquAdvantage salmon can mature in half the time as wild salmon. AquaBounty applied for regulatory approval to market their GM salmon in the US, and was approved in November 2015. On 25 November 2013 Canada approved commercial scale production and export of GM Salmon eggs but they are not approved for human consumption in Canada.

Several academic groups have been developing GM zebrafish to detect aquatic pollution. The lab that originated the GloFish discussed above originally developed them to change color in the presence of pollutants, to be used as environmental sensors. A lab at University of Cincinnati has been developing GM zebrafish for the same purpose, as has a lab at Tulane University.

Recent research on pain in fish has resulted in concerns being raised that genetic-modifications induced for scientific research may have detrimental effects on the welfare of fish.

Frogs

Genetically modified frogs are used for scientific research and are widely used in basic research including genetics and early development. Two species of frog, *Xenopus laevis* and *Xenopus tropicalis*, are most commonly used.

GM frogs are also being used as pollution sensors, especially for endocrine disrupting chemicals.

Invertebrates

Fruit Flies

In biological research, transgenic fruit flies (*Drosophila melanogaster*) are model organisms used to study the effects of genetic changes on development. Fruit flies are often preferred over other animals due to their short life cycle, low maintenance requirements, and relatively simple genome compared to many vertebrates.

Mosquitoes

In 2010, scientists created "malaria-resistant mosquitoes" in the laboratory. The World Health Organization estimated that malaria killed almost one million people in 2008. Genetically modified male mosquitoes containing a lethal gene have been developed to combat the spread of dengue fever and the Zika virus. *Aedes aegypti* mosquitoes, the single most important carrier of dengue fever and the Zika virus, were reduced by 80% in a 2010 trial of these GM mosquitoes in the Cayman Islands and by 90% in a 2015 trial in Bahia, Brazil. In comparison, the Florida Keys Mosquito Control District has achieved only 30%-60% population reduction with traps and pesticide spraying. In 2016 FDA approved a genetically modified mosquito intervention for Key West, Florida. UK firm Oxitec proposed the release of millions of modified male (non-biting) mosquitoes to compete with wild males for mates. The males are engineered so that their offspring die before maturing, helping to eradicate mosquito-borne disease. Final approval was to be based on a local referendum to be held in November. Andrea Crisanti, a molecular biologist at Imperial College in London is working on ways to stop the A. gambiae mosquito from transmitting disease.

Bollworms

A strain of *Pectinophora gossypiella* (Pink bollworm) has been genetically engineered to express a red fluorescent protein. This allows researchers to monitor bollworms that have been sterilized by radiation and released to reduce bollworm infestation. The strain has been field tested for over three years and has been approved for release.

Cnidaria

Cnidaria such as *Hydra* and the sea anemone *Nematostella vectensis* are attractive model organisms to study the evolution of immunity and certain developmental processes. An important technical breakthrough was the development of procedures for generation of stable transgenic hydras and sea anemones by embryo microinjection.

Regulation

The regulation of genetic engineering concerns the approaches taken by governments to assess and manage the risks associated with the use of genetic engineering technology and the development and release of genetically modified organisms (GMO), including genetically modified crops and genetically modified fish. There are differences in the regulation of GMOs between countries, with some of the most marked differences occurring between the USA and Europe. Regulation varies in a given country depending on the intended use of the products of the genetic engineering. For example, a crop not intended for food use is generally not reviewed by authorities responsible for food safety. The European Union differentiates between approval for cultivation within the EU and approval for import and processing. While only a few GMOs have been approved for cultivation in the EU a number of GMOs have been approved for import and processing. The cultivation of GMOs has triggered a debate about the market for GMOs in Europe. Depending on the coexistence regulations, incentives for cultivation of GM crops differ.

Controversy

There is controversy over GMOs, especially with regard to their use in producing food.

The dispute involves buyers, biotechnology companies, governmental regulators, non-governmental organizations, and scientists. The key areas of controversy related to GMO food are whether GM food should be labeled, the role of government regulators, the effect of GM crops on health and the environment, the effect on pesticide resistance, the impact of GM crops for farmers, and the role of GM crops in feeding the world population. In 2014, sales of products that had been labeled as non-GMO grew 30 percent to $1.1 billion.

There is a scientific consensus that currently available food derived from GM crops poses no greater risk to human health than conventional food, but that each GM food needs to be tested on a case-by-case basis before introduction. Nonetheless, members of the public are much less likely than scientists to perceive GM foods as safe. The legal and regulatory status of GM foods varies by country, with some nations banning or restricting them, and others permitting them with widely differing degrees of regulation.

No reports of ill effects have been proven in the human population from ingesting GM food. Although labeling of GMO products in the marketplace is required in many countries, it is not required in the United States and no distinction between marketed GMO and non-GMO foods is recognized by the US FDA. In an article in *The Economist* it was argued that, while GM foods could potentially help feed 842 million malnourished people globally, laws such as the one passed in Vermont, to require labeling of foods containing genetically modified ingredients, could have the unintended consequence of interrupting the process of spreading GM technologies to impoverished countries that suffer with food security problems.

A 2014 critical review of histopathology studies on rats (eating approved widely eaten GM crops) found significant flaws, inadequacies, and a lack of transparency in methodology and results. Published studies could be found for only 19% of these widely eaten crops. Most of reviewed studies were performed after the approval of crop. More research is needed, including long-term animal feeding studies and thorough histopathological investigations.

The Organic Consumers Association, and the Union of Concerned Scientists, and Greenpeace stated that risks have not been adequately identified and managed, and they have questioned the objectivity of regulatory authorities. Some health groups say there are unanswered questions regarding the potential long-term impact on human health from food derived from GMOs, and propose mandatory labeling or a moratorium on such products. Concerns include contamination of the non-genetically modified food supply, effects of GMOs on the environment and nature, the rigor of the regulatory process, and consolidation of control of the food supply in companies that make and sell GMOs, or concerns over the use of herbicides with glyphosate.

Genetically Modified Virus

A genetically modified virus is a virus that has gone through genetic modification for various biomedical purposes.

General usage

Genetic modification involves the insertion or deletion of genes. When genes are inserted, they usually come from different species, which is a form of horizontal gene transfer. In nature this can occur when exogenous DNA penetrates the cell membrane for any reason but usually for domination of other diseases.

To do this artificially may require attaching the genes to a virus or just physically inserting the extra DNA into the nucleus of the intended host with a very small syringe, or with very small particles fired from a gene gun. However, other methods exploit natural forms of gene transfer, such as the ability of *Agrobacterium* to transfer genetic material to plants, or the ability of lentiviruses to transfer genes to animal cells.

Lithium-ion Batteries

In materials science, a genetically modified virus has been used to construct a more environmentally friendly lithium-ion battery.

Gene Therapy

Gene therapy uses genetically modified viruses to deliver genes that can cure disease into human cells. Although gene therapy is still relatively new, it has had some successes. It has been used to treat genetic disorders such as severe combined immunodeficiency.

Heart Pacemaker

In 2012, US researchers reported that they injected a genetically modified virus into the heart of guinea pigs. This virus inserted into the heart muscles a gene called Tbx18 which enabled heartbeats. The researchers forecast that one day this technique could be used to restore the heartbeat in humans who would otherwise need electronic pacemakers.

Cancer Treatment

In 2004, researchers reported that a genetically modified virus that exploits the selfish behaviour of cancer cells might offer an alternative way of killing tumours. Since then, several researchers have developed genetically modified oncolytic viruses that show promise as treatments for various types of cancer

Rabbits

In Spain and Portugal, by 2005 rabbits had declined by as much as 95% over 50 years due diseases such as myxomatosis, rabbit haemorrhagic disease and other causes. This in turn caused declines in predators like the Iberian lynx, a critically endangered species. In 2000 Spanish researchers investigated a genetically modified virus which might have protected rabbits in the wild against myxomatosis and rabbit haemorrhagic disease. However, there was concern that such a virus might make its way into wild populations in areas such as Australia and create a population boom. Rabbits in Australia are considered to be such a pest that land owners are legally obliged to control them.

GMO Lentivirus

A scientist claims she was infected by a genetically modified virus while working for Pfizer. In her federal lawsuit she says she has been intermittently paralyzed by the Pfizer-designed virus. "McClain, of Deep River, suspects she was inadvertently exposed, through work by a former Pfizer colleague in 2002 or 2003, to an engineered form of the lentivirus, a virus similar to the one that can lead to acquired immune deficiency syndrome, or AIDS." The court found that McClain failed to demonstrate that her illness was caused by exposure to the lentivirus, but also that Pfizer violated whistleblower laws.

Biohazard Research Limitations

The National Institute of Health declared a research funding moratorium on select Gain-of-Function virus research in January 2015. Questions about a potential escape of a modified virus from a biosafety lab and the utility of dual-use-technology, dual use research of concern (DURC), prompted the NIH funding policy revision.

Various Techniques

Gene Knockdown

Gene knockdown is an experimental technique by which the expression of one or more of an organism's genes are reduced. The reduction can occur either through genetic modification or by treatment with a reagent such as a short DNA or RNA oligonucleotide that has a sequence complementary to either gene or an mRNA transcript.

Versus Transient Knockdown

If genetic modification of DNA is done, the result is called "knockdown organism." If the change in gene expression is caused by an oligonucleotide binding to an mRNA or temporarily binding to a gene, this leads to a temporary change in gene expression that does not modify the chromosomal DNA, and the result is referred to as a "transient knockdown".

In a transient knockdown, the binding of this oligonucleotide to the active gene or its transcripts causes decreased expression through a variety of processes. Binding can occur either through the blocking of transcription (in the case of gene-binding), the degradation of the mRNA transcript (e.g. by small interfering RNA (siRNA)) or RNase-H dependent antisense), or through the blocking of either mRNA translation, pre-mRNA splicing sites, or nuclease cleavage sites used for maturation of other functional RNAs, including miRNA (e.g. by morpholino oligos or other RNase-H independent antisense).

The most direct use of transient knockdowns is for learning about a gene that has been sequenced, but has an unknown or incompletely known function. This experimental approach is known as reverse genetics. Researchers draw inferences from how the knockdown differs from individuals in which the gene of interest is operational. Transient knockdowns are often used in developmental biology because oligos can be injected into single-celled zygotes and will be present in the daughter cells of the injected cell through embryonic development.

RNA Interference

RNA interference (RNAi) is a means of silencing genes by way of mRNA degradation. Gene knockdown by this method is achieved by introducing small double-stranded interfering RNAs (siRNA) into the cytoplasm. Small interfering RNAs can originate from inside the cell or can be exogenously introduced into the cell. Once introduced into the cell, exogenous siRNAs are processed by the RNA-induced silencing complex (RISC). The siRNA is complementary to the target mRNA to be silenced, and the RISC uses the siRNA as a template for locating the target mRNA. After the RISC localizes to the target mRNA, the RNA is cleaved by a ribonuclease.

RNAi is widely used as a laboratory technique for genetic functional analysis. RNAi in organisms such as *C. elegans* and *Drosophila melanogaster* provides a quick and inexpensive means of investigating gene function. In *C. elegans* research, the availability of tools such as the Ahringer RNAi Library give laboratories a way of testing many genes in a variety of experimental backgrounds. Insights gained from experimental RNAi use may be useful in identifying potential therapeutic targets, drug development, or other applications. RNA interference is a very useful research tool, allowing investigators to carry out large genetic screens in an effort to identify targets for further research related to a particular pathway, drug, or phenotype.

CRISPRs

A different means of silencing exogenous DNA that has been discovered in prokaryotes is a mechanism involving loci called 'Clustered Regularly Interspaced Short Palindromic Repeats', or CRISPRs. Proteins called 'CRISPR-associated genes' (cas genes) encode cellular machinery that cuts exogenous DNA into small fragments and inserts them into a CRISPR repeat locus. When this CRISPR region of DNA is expressed by the cell, the small RNAs produced from the exogenous DNA inserts serve as a template sequence that other Cas proteins use to silence this same exogenous sequence. The transcripts of the short exogenous sequences are used as a guide to silence these foreign DNA when they are present in the cell. This serves as a kind of acquired immunity, and this process is like a prokaryotic RNA interference mechanism. The CRISPR repeats are conserved amongst many species and have been demonstrated to be usable in human cells, bacteria, *C. elegans*, zebrafish, and other organisms for effective genome manipulation. The use of CRISPRs as a versatile research tool can be illustrated by many studies making use of it to generate organisms with genome alterations.

TALENs

Another technology made possible by prokaryotic genome manipulation is the use of transcription activator-like effector nucleases (TALENs) to target specific genes. TALENs are nucleases that have two important functional components: a DNA binding domain and a DNA cleaving domain. The DNA binding domain is a sequence-specific transcription activator-like effector sequence while the DNA cleaving domain originates from a bacterial endonuclease and is non-specific. TALENs can be designed to cleave a sequence specified by the sequence of the transcription activator-like effector portion of the construct. Once designed, a TALEN is introduced into a cell as a plasmid or mRNA. The TALEN is expressed, localizes to its target sequence, and cleaves a specific site. After cleavage of the target DNA sequence by the TALEN, the cell uses non-homologous end joining as a DNA repair mechanism to correct the cleavage. The cell's attempt at repairing the cleaved sequence can render the encoded protein non-functional, as this repair mechanism introduces insertion or deletion errors at the repaired site.

Commercialization

So far, knockdown organisms with permanent alterations in their DNA have been engineered chiefly for research purposes. Also known simply as knockdowns, these organisms are most commonly used for reverse genetics, especially in species such as mice or rats for which transient knockdown technologies cannot easily be applied.

There are several companies that offer commercial services related to gene knockdown treatments.

Gene Knockout

A gene knockout (abbreviation: KO) is a genetic technique in which one of an organism's genes is made inoperative ("knocked out" of the organism). Also known as knockout organisms or simply knockouts, they are used in learning about a gene that has been sequenced, but which has an unknown or incompletely known function. Researchers draw inferences from the difference between the knockout organism and normal individuals.

The term also refers to the process of creating such an organism, as in "knocking out" a gene. The technique is essentially the opposite of a gene knockin. Knocking out two genes simultaneously in an organism is known as a double knockout (DKO). Similarly the terms triple knockout (TKO) and quadruple knockouts (QKO) are used to describe three or four knocked out genes, respectively.

Method

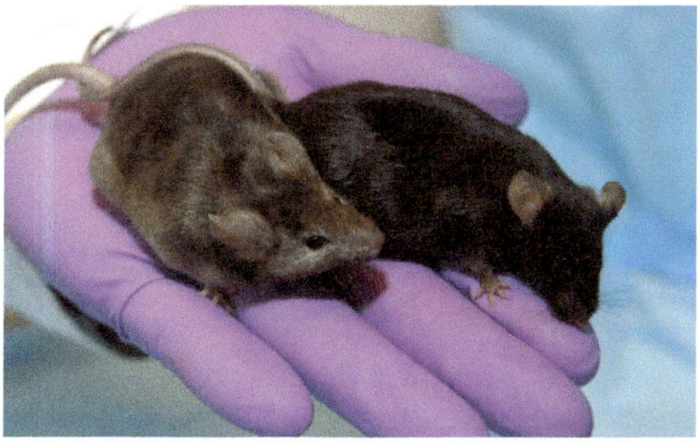

A laboratory mouse in which a gene affecting hair growth has been knocked out (left), is shown next to a normal lab mouse.

Knockout is accomplished through a combination of techniques, beginning in the test tube with a plasmid, a bacterial artificial chromosome or other DNA construct, and proceeding to cell culture. Individual cells are genetically transfected with the DNA construct. Often the goal is to create a transgenic animal that has the altered gene. If so, embryonic stem cells are genetically transformed and inserted into early embryos. Resulting animals with the genetic change in their germline cells can then often pass the gene knockout to future generations.

To create knockout moss, transfection of protoplasts is the preferred method. Such transformed *Physcomitrella*-protoplasts directly regenerate into fertile moss plants. Eight weeks after transfection, the plants can be screened for gene targeting via PCR.

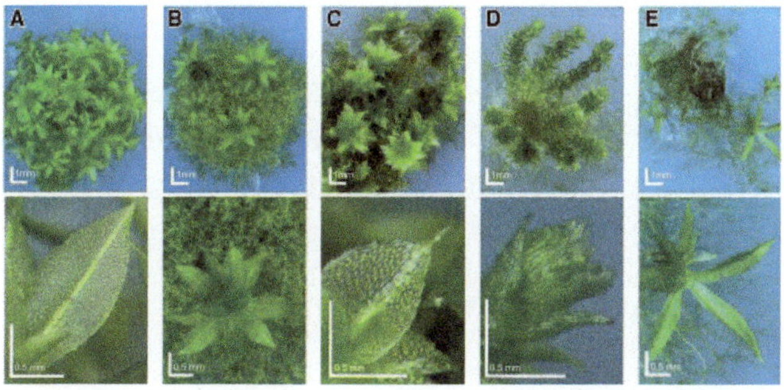

Wild-type Physcomitrella and knockout mosses: Deviating phenotypes induced in gene-disruption library transformants. Physcomitrella wild-type and transformed plants were grown on minimal Knop medium to induce differentiation and development of gametophores. For each plant, an overview (upper row; scale bar corresponds to 1 mm) and a close-up (bottom row; scale bar equals 0.5 mm) are shown. A: Haploid wild-type moss plant completely covered with leafy gametophores and close-up of wild-type leaf. B–D: Different mutants.

The construct is engineered to recombine with the target gene, which is accomplished by incorporating sequences from the gene itself into the construct. Recombination then occurs in the region of that sequence within the gene, resulting in the insertion of a foreign sequence to disrupt the gene. With its sequence interrupted, the altered gene in most cases will be translated into a nonfunctional protein, if it is translated at all.

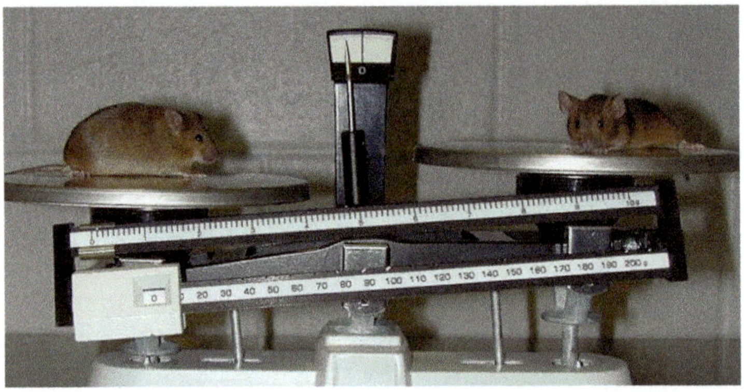

A knockout mouse (left) that is a model of obesity, compared with a normal mouse.

A conditional knockout allows gene deletion in a tissue or time specific manner. This is done by introducing short sequences called loxP sites around the gene. These sequences will be introduced into the germ-line via the same mechanism as a knock-out. This germ-line can then be crossed to another germline containing Cre-recombinase which is a viral enzyme that can recognize these sequences, recombines them and deletes the gene flanked by these sites.

Because the desired type of DNA recombination is a rare event in the case of most cells and most constructs, the foreign sequence chosen for insertion usually includes a reporter. This enables easy selection of cells or individuals in which knockout was successful. Sometimes the DNA construct inserts into a chromosome without the desired homologous recombination with the target gene. To eliminate such cells, the DNA construct often contains a second region of DNA that allows such cells to be identified and discarded.

In diploid organisms, which contain two alleles for most genes, and may as well contain several related genes that collaborate in the same role, additional rounds of transformation and selection are performed until every targeted gene is knocked out. Selective breeding may be required to produce homozygous knockout animals.

Gene knockin is similar to gene knockout, but it replaces a gene with another instead of deleting it.

Use

Knockouts are primarily used to understand the role of a specific gene or DNA region by comparing the knockout organism to a wildtype with a similar genetic background.

Knockouts organisms are also used as screening tools in the development of drugs, to target specific biological processes or deficiencies by using a specific knockout, or to understand the mechanism of action of a drug by using a library of knockout organisms spanning the entire genome, such as in *Saccharomyces cerevisiae*.

Gene Targeting

Gene targeting (also, replacement strategy based on homologous recombination) is a genetic technique that uses homologous recombination to change an endogenous gene. The method can be used to delete a gene, remove exons, add a gene, and introduce point mutations. Gene targeting can be permanent or conditional. Conditions can be a specific time during development / life of the organism or limitation to a specific tissue, for example. Gene targeting requires the creation of a specific vector for each gene of interest. However, it can be used for any gene, regardless of transcriptional activity or gene size.

Methods

Gene targeting methods are established for several model organisms and may vary depending on the species used. In general, a targeting construct made out of DNA is generated in bacteria. It typically contains part of the gene to be targeted, a reporter gene, and a (dominant) selectable marker.

To target genes in mice, this construct is then inserted into mouse embryonic stem cells in culture. After cells with the correct insertion have been selected, they can be used to contribute to a mouse's tissue via embryo injection. Finally, chimeric mice where the modified cells made up the reproductive organs are selected for via breeding. After this step the entire body of the mouse is based on the previously selected embryonic stem cell.

To target genes in moss, this construct is incubated together with freshly isolated protoplasts and with Polyethylene glycol. As mosses are haploid organisms, regenerating moss filaments (protonema) can directly be screened for gene targeting, either by treatment with antibiotics or with PCR. Unique among plants, this procedure for reverse genetics is as efficient as in yeast. Using modified procedures, gene targeting has also been successfully applied to cattle, sheep, swine, and many fungi.

The frequency of gene targeting can be significantly enhanced through the use of engineered endonucleases such as zinc finger nucleases, engineered homing endonucleases, and nucleases based on engineered TAL effectors. To date, this method has been applied to a number of species including Drosophila melanogaster, tobacco, corn, human cells, mice, and rats.

Comparison with Gene Trapping

Gene trapping is based on random insertion of a cassette while gene targeting targets a specific gene. Cassettes can be used for many different things while the flanking homology regions of gene targeting cassettes need to be adapted for each gene. This makes gene trapping more easily amenable for large scale projects than targeting. On the other hand, gene targeting can be used for genes with low transcriptions that would go undetected in a trap screen. Also, the probability of trapping increases with intron size. For gene targeting these compact genes are just as easily altered.

Applications

Gene targeting has been widely used to study human genetic diseases by removing ("knocking out"), or adding ("knocking in"), specific mutations of interest to a variety of models. Previously used to engineer rat cell models, advances in gene targeting technologies are enabling the creation of a new wave of isogenic human disease models. These models are the most accurate in-vitro models available to researchers to date, and are facilitating the development of new personalized drugs and diagnostics, particularly in the field of cancer.

2007 Nobel Prize

Mario R. Capecchi, Martin J. Evans and Oliver Smithies were declared laureates of the 2007 Nobel Prize in Physiology or Medicine for their work on "principles for introducing specific gene modifications in mice by the use of embryonic stem cells", or gene targeting.

Types of Transgenic Animals

Transgenic animals are genetically altered with specific characteristics which otherwise would not be present in that specific animal. In general, transgenic animals have either DNA added (to express an additional gene) or have their genome altered (to abolish or modify the expression of an existing gene). Rodents particularly mice comprise of over 95% of transgenic animals used in biomedical research. The mouse is the model organism of choice because of the following reasons:

- Complete mice genome sequence is available.
- Easy genetic manipulation of mice cells and embryos.
- Short gestation period and large litter size.
- Availability of major antibodies and other molecular biology tools.
- Possibility to perform physiologic and behavioral tests that can be directly linked to human disease.

Other transgenic species include cattle, pig, sheep and rats. Their use in pharmaceutical research has so far been limited due to technical constraints. Recent advances in molecular biology techniques may allow us to use transgenic rat for the development of many human therapeutics where the rat is a better model than the mouse.

Uses of Transgenic Animals

Transgenic animals are useful in the discovery of new therapy for important human diseases.

Transgenic animals are fundamentally similar to their counter wild type variety except for some genomic heterogeneity.

Transgenic animals are used in the gene therapy experiment to understand the importance of malfunctioned or mutated gene.

Transgenic animals are also used to check the efficacy and safety of the drugs or vaccines used in the clinical trials.

Transgenic mice can be generated to express human targeted gene that can be further used to design new therapy. Transgenic mice can also obviate the use of animals such as monkeys for testing drugs for many human diseases, eg. hypercholesterolemia and HIV.

Some terminology:

The number of offsprings produced at one birth by an animal is called Litter size.

Normal pregnancy period is called as gestation period.

Presence of high levels of cholesterol in the blood is called as hypercholesterolemia.

LDL receptor (LDLR) present in liver helps to eliminate the excess cholesterol from the body. The deficiency of LDLR in human body leads to increase in blood cholesterol level, condition called as HYPERCHOLESTEREMIA.

n hypercholesteremia chances of getting heart attack is higher because of the deposition of cholesterol in the lining of blood vessels. Blood plasma is purified for any unwanted substance in order to prolong the life of a patient (eg low density lipoprotein [LDL] from blood) by a method called PLASMAPHERESIS.

The historical example of a transgenic animal goes to successful generation of a mice where the mouse gene for metallothionein-I was fused to the human growth hormone (GH) and introduced into mice. The resultant transgenic mice showed altered growth characteristics and served as a valuable resource for human disease condition where production of excess growth hormone modulates many physiological processes.

Generation of Transgenic Animal

Transgenic animals can be generated by following methods:

- Retroviral infection of pre- or pro-implantation embryos
- DNA injection of embryos at pronuclear stage.
- Microinjection of genetically modified embryonic stem cells into blastocysts.

The transgenic animals are created mostly by a well- known technique where fertilized embryos were microinjected by plasmid DNA containing a fused protein *in vitro* . A micro needle is used to inject the targeted plasmid DNA into the embryo that leads to integration of foreign DNA into the host genome. The pronuclear embryo containing the foreign DNA is then implanted into the recipient animal.

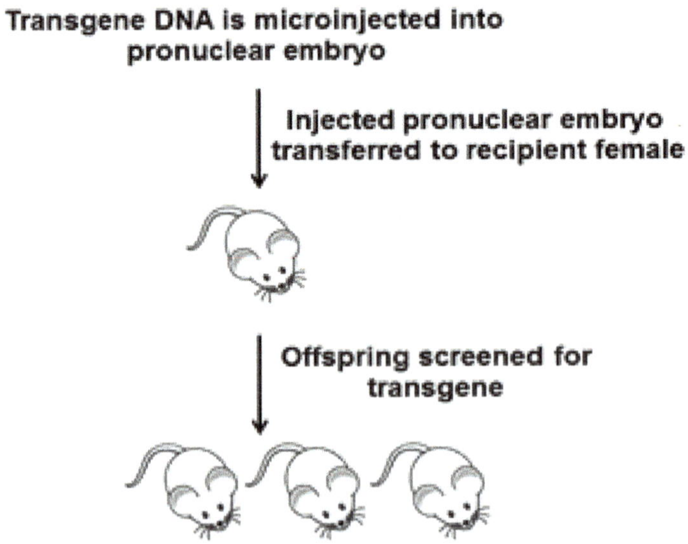

Standard transgenic approach

In another approach, the pluripotent stem cells derived from embryonic blastocysts are elctroporated by foreign DNA containing the gene of interest. The microinjected stem cells are reintroduced into the blastocysts which are then transferred into the uterus of a pseudopregnant recipient animal.

Both the methods are successful only after multiple generations of breeding and selection of the most stable transgenic line. Plasmid DNA containing gene of interest can be coupled with a tissue specific promoter to make it as an inducible system. Cytomegalovirus promoter is used widely as a promoter for the inducible system. Alternatively the recombination of gene of interest can also be achieved by *Cre-lox* pathway.

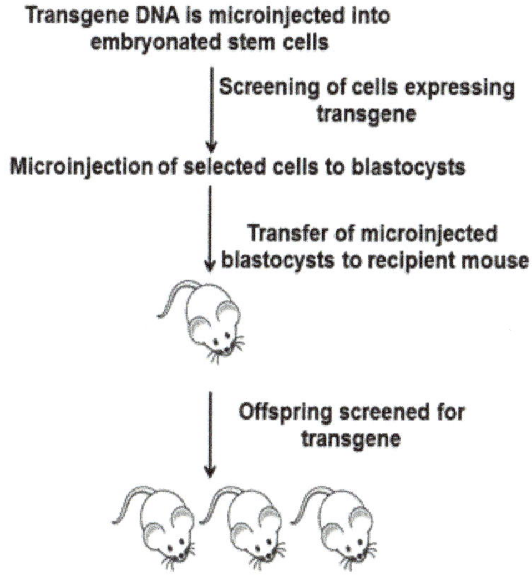

Alternate approach

Genetic Recombination

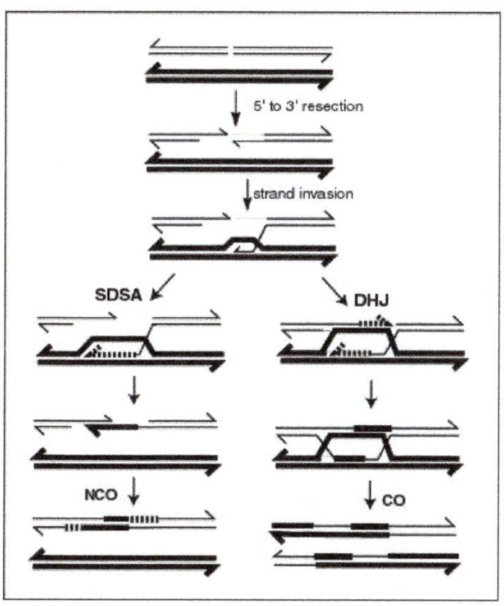

A current model of meiotic recombination, initiated by a double-strand break or gap, followed by pairing with an homologous chromosome and strand invasion to initiate the recombinational repair process. Repair of the gap can lead to crossover (CO) or non-crossover (NCO) of the flanking regions. CO recombination is thought to occur by the Double Holliday Junction (DHJ) model, illustrated on the right, above. NCO recombinants are thought to occur primarily by the Synthesis Dependent Strand Annealing (SDSA) model, illustrated on the left, above. Most recombination events appear to be the SDSA type.

Genetic recombination is the production of offspring with combinations of traits that differ from those found in either parent. In eukaryotes, genetic recombination during meiosis can lead to a novel set of genetic information that can be passed on from the parents to the offspring. Most recombination is naturally occurring.

During meiosis in eukaryotes, genetic recombination involves the pairing of homologous chromosomes. This may be followed by information transfer between the chromosomes. The information transfer may occur without physical exchange (a section of genetic material is copied from one chromosome to another, without the donating chromosome being changed) ; or by the breaking and rejoining of DNA strands, which forms new molecules of DNA.

Recombination may also occur during mitosis in eukaryotes where it ordinarily involves the two sister chromosomes formed after chromosomal replication. In this case, new combinations of alleles are not produced since the sister chromosomes are usually identical. In meiosis and mitosis, recombination occurs between similar molecules of DNA (homologs). In meiosis, non-sister homologous chromosomes pair with each other so that recombination characteristically occurs between non-sister homologues. In both meiotic and mitotic cells, recombination between homologous chromosomes is a common mechanism used in DNA repair.

Genetic recombination and recombinational DNA repair also occurs in bacteria and archaea, which use asexual reproduction.

Recombination can be artificially induced in laboratory (in vitro) settings, producing recombinant DNA for purposes including vaccine development.

V(D)J recombination in organisms with an adaptive immune system is a type of site-specific genetic recombination that helps immune cells rapidly diversify to recognize and adapt to new pathogens.

Synapsis

During meiosis, synapsis (the pairing of homologous chromosomes) ordinarily precedes genetic recombination.

Mechanism

Genetic recombination is catalyzed by many different enzymes. Recombinases are key enzymes that catalyse the strand transfer step during recombination. RecA, the chief recombinase found in *Escherichia coli*, is responsible for the repair of DNA double strand breaks (DSBs). In yeast and other eukaryotic organisms there are two recombinases required for repairing DSBs. The RAD51 protein is required for mitotic and meiotic recombination, whereas the DNA repair protein, DMC1, is specific to meiotic recombination. In the archaea, the ortholog of the bacterial RecA protein is RadA.

Bacterial recombination

In Bacteria there are:

- regular bacterial recombination, as well as noneffective transfer of genetic material, expressed as
- unsuccessful transfer or abortive transfer which is any bacterial DNA transfer of the donor cell recipients who have set the incoming DNA as part of the genetic material of the recipient. Abortive transfer was registered in the following transduction and conjugation. In all cases, the transmitted fragment is diluted by the culture growth.

Chromosomal Crossover

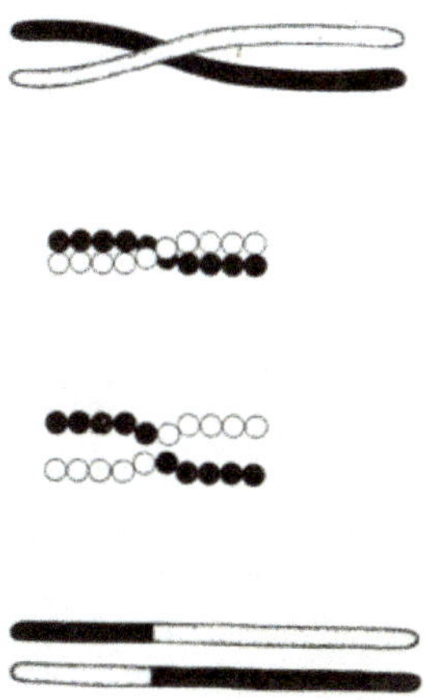

Thomas Hunt Morgan's illustration of crossing over (1916)

In eukaryotes, recombination during meiosis is facilitated by chromosomal crossover. The crossover process leads to offspring having different combinations of genes from those of their parents, and can occasionally produce new chimeric alleles. The shuffling of genes brought about by genetic recombination produces increased genetic variation. It also allows sexually reproducing organisms to avoid Muller's ratchet, in which the genomes of an asexual population accumulate genetic deletions in an irreversible manner.

Chromosomal crossover involves recombination between the paired chromosomes inherited from each of one's parents, generally occurring during meiosis. During prophase I (pachytene stage) the four available chromatids are in tight formation with one another. While in this formation, homologous sites on two chromatids can closely pair with one another, and may exchange genetic information.

Because recombination can occur with small probability at any location along chromosome, the frequency of recombination between two locations depends on the distance separating them. Therefore, for genes sufficiently distant on the same chromosome the amount of crossover is high enough to destroy the correlation between alleles.

Tracking the movement of genes resulting from crossovers has proven quite useful to geneticists. Because two genes that are close together are less likely to become separated than genes that are farther apart, geneticists can deduce roughly how far apart two genes are on a chromosome if they know the frequency of the crossovers. Geneticists can also use this method to infer the presence of certain genes. Genes that typically stay together during recombination are said to be linked. One gene in a linked pair can sometimes be used as a marker to deduce the presence of another gene. This is typically used in order to detect the presence of a disease-causing gene.

The recombination frequency between two loci observed is the *crossing-over value*. It is the frequency of crossing over between two linked gene loci (markers), and depends on the mutual distance of the genetic loci observed. For any fixed set of genetic and environmental conditions, recombination in a particular region of a linkage structure (chromosome) tends to be constant, and the same is then true for the crossing-over value which is used in the production of genetic maps.

Gene Conversion

In gene conversion, a section of genetic material is copied from one chromosome to another, without the donating chromosome being changed. Gene conversion occurs at high frequency at the actual site of the recombination event during meiosis. It is a process by which a DNA sequence is copied from one DNA helix (which remains unchanged) to another DNA helix, whose sequence is altered. Gene conversion has often been studied in fungal crosses where the 4 products of individual meioses can be conveniently observed. Gene conversion events can be distinguished as deviations in an individual meiosis from the normal 2:2 segregation pattern (e.g. a 3:1 pattern).

Nonhomologous Recombination

Recombination can occur between DNA sequences that contain no sequence homology. This can cause chromosomal translocations, sometimes leading to cancer.

In B cells

B cells of the immune system perform genetic recombination, called immunoglobulin class switching. It is a biological mechanism that changes an antibody from one class to another, for example, from an isotype called IgM to an isotype called IgG.

Genetic Engineering

In genetic engineering, recombination can also refer to artificial and deliberate recombination of disparate pieces of DNA, often from different organisms, creating what is called recombinant DNA. A prime example of such a use of genetic recombination is gene targeting, which can be used to add, delete or otherwise change an organism's genes. This technique is important to biomedical researchers as it allows them to study the effects of specific genes. Techniques based on genetic recombination are also applied in protein engineering to develop new proteins of biological interest.

Recombinational Repair

During both mitosis and meiosis, DNA damages caused by a variety of exogenous agents (e.g. UV light, X-rays, chemical cross-linking agents) can be repaired by homologous recombinational repair (HRR). These findings suggest that DNA damages arising from natural processes, such as exposure to reactive oxygen species that are byproducts of normal metabolism, are also repaired by HRR. In humans and rodents, deficiencies in the gene products necessary for HRR during meiosis cause infertility. In humans, deficiencies in gene products necessary for HRR, such as BRCA1 and BRCA2, increase the risk of cancer.

In bacteria, transformation is a process of gene transfer that ordinarily occurs between individual cells of the same bacterial species. Transformation involves integration of donor DNA into the recipient chromosome by recombination. This process appears to be an adaptation for repairing DNA damages in the recipient chromosome by HRR. Transformation may provide a benefit to pathogenic bacteria by allowing repair of DNA damage, particularly damages that occur in the inflammatory, oxidizing environment associated with infection of a host.

When two or more viruses, each containing lethal genomic damages, infect the same host cell, the virus genomes can often pair with each other and undergo HRR to produce viable progeny. This process, referred to as multiplicity reactivation, has been studied in lambda and T4 bacteriophages, as well as in several pathogenic viruses. In the case of pathogenic viruses, multiplicity reactivation may be an adaptive benefit to the virus since it allows the repair of DNA damages caused by exposure to the oxidizing environment produced during host infection.

Meiotic Recombination

Molecular models of meiotic recombination have evolved over the years as relevant evidence accumulated. A major incentive for developing a fundamental understanding of the mechanism of meiotic recombination is that such understanding is crucial for solving the problem of the adaptive function of sex, a major unresolved issue in biology. A recent model that reflects current understanding was presented by Anderson and Sekelsky. The figure shows that two of the four chromatids present early in meiosis (prophase I) are paired with each other and able to interact. Recombination, in this version of the model, is initiated by a double-strand break (or gap) shown in the DNA molecule (chromatid) at the top of the first figure in this article. However, other types of

DNA damage may also initiate recombination. For instance, an inter-strand cross-link (caused by exposure to a cross-linking agent such as mitomycin C) can be repaired by HRR.

As indicated in the first figure, above, two types of recombinant product are produced. Indicated on the right side is a "crossover" (CO) type, where the flanking regions of the chromosomes are exchanged, and on the left side, a "non-crossover" (NCO) type where the flanking regions are not exchanged. The CO type of recombination involves the intermediate formation of two "Holliday junctions" indicated in the lower right of the figure by two X shaped structures in each of which there is an exchange of single strands between the two participating chromatids. This pathway is labeled in the figure as the DHJ (double-Holliday junction) pathway.

The NCO recombinants are produced by a process referred to as "synthesis dependent strand annealing" (SDSA). Recombination events of the NCO/SDSA type appear to be more common than the CO/DHJ type. The NCO/SDSA pathway contributes little to genetic variation since the arms of the chromosomes flanking the recombination event remain in the parental configuration. Thus, explanations for the adaptive function of meiosis that focus exclusively on crossing-over are inadequate to explain the majority of recombination events.

Achiasmy and Heterochiasmy

Achiasmy is the phenomenon where autosomal recombination is completely absent in one sex of a species. Achiasmatic chromosomal segregation is well documented in male *Drosophila melanogaster*. Heterochiasmy is the term used to describe recombination rates which differ between the sexes of a species. This sexual dimorphic pattern in recombination rate has been observed in many species. In mammals, females most often have higher rates of recombination. The "Haldane-Huxley rule" states that achiasmy usually occurs in the heterogametic sex.

Artificial Gene Synthesis

Artificial gene synthesis, sometimes known as DNA printing is a method in synthetic biology that is used to create artificial genes in the laboratory. Currently based on solid-phase DNA synthesis, it differs from molecular cloning and polymerase chain reaction (PCR) in that the user does not have to begin with preexisting DNA sequences. Therefore, it is possible to make a completely synthetic double-stranded DNA molecule with no apparent limits on either nucleotide sequence or size. The method has been used to generate functional bacterial or yeast chromosomes containing approximately one million base pairs. Recent research also suggests the possibility of creating novel nucleobase pairs in addition to the two base pairs in nature, which could greatly expand the genetic code.

Synthesis of the first complete gene, a yeast tRNA, was demonstrated by Har Gobind Khorana and coworkers in 1972. Synthesis of the first peptide- and protein-coding genes was performed in the laboratories of Herbert Boyer and Alexander Markham, respectively.

Commercial gene synthesis services are now available from numerous companies worldwide, some of which have built their business model around this task. Current gene synthesis approaches are most often based on a combination of organic chemistry and molecular biological techniques and entire genes may be synthesized "de novo", without the need for precursor template DNA. Gene synthesis has become an important tool in many fields of recombinant DNA technology including heterologous gene expression, vaccine development, gene therapy and molecular engineering. The synthesis of nucleic acid sequences can be more economical than classical cloning and mutagenesis procedures in addition to being a powerful and flexible engineering tool for creating and designing new DNA sequences and protein functions.

Gene Optimization

While the ability to make increasingly long stretches of DNA efficiently and at lower prices is a technological driver of this field, increasingly attention is being focused on improving the design of genes for specific purposes. Early in the genome sequencing era, gene synthesis was used as an (expensive) source of cDNAs that were predicted by genomic or partial cDNA information but were difficult to clone. As higher quality sources of sequence verified cloned cDNA have become available, this practice has become less urgent.

Producing large amounts of protein from gene sequences (or at least the protein coding regions of genes, the open reading frame) found in nature can sometimes prove difficult and is a problem of sufficient impact that scientific conferences have been devoted to the topic. Many of the most interesting proteins sought by molecular biologists are normally regulated to be expressed in very low amounts in wild type cells. Redesigning these genes offers a means to improve gene expression in many cases. Rewriting the open reading frame is possible because of the degeneracy of the genetic code. Thus it is possible to change up to about a third of the nucleotides in an open reading frame and still produce the same protein. The available number of alternate designs possible for a given protein is astronomical. For a typical protein sequence of 300 amino acids there are over 10^{150} codon combinations that will encode an identical protein. Using optimization methods such as replacing rarely used codons with more common codons sometimes have dramatic effects. Further optimizations such as removing RNA secondary structures can also be included. At least in the case of *E. coli*, protein expression is maximized by predominantly using codons corresponding to tRNA that retain amino acid charging during starvation. Computer programs written to perform these, and other simultaneous optimizations are used to handle the enormous complexity of the task. A well optimized gene can improve protein expression 2 to 10 fold, and in some cases more than 100 fold improvements have been reported. Because of the large numbers of nucleotide changes made to the original DNA sequence, the only practical way to create the newly designed genes is to use gene synthesis.

Standard Methods

Oligonucleotide Synthesis

Oligonucleotides are chemically synthesized using building blocks called nucleoside phosphoramidites. These can be normal or modified nucleosides which have protecting groups to prevent their amines, hydroxyl groups and phosphate groups from interacting incorrectly. One phosphoramidite is added at a time, the 5' hydroxyl group is deprotected and a new base is added and so on. The chain grows in the 3' to 5' direction, which is backwards relative to biosynthesis. At the end, all the protecting groups are removed. Nevertheless, being a chemical process, several incorrect interactions occur leading to some defective products. The longer the oligonucleotide sequence that is being synthesized, the more defects there are, thus this process is only practical for producing short sequences of nucleotides. The current practical limit is about 200 bp (base pairs) for an oligonucleotide with sufficient quality to be used directly for a biological application. HPLC can be used to isolate products with the proper sequence. Meanwhile, a large number of oligos can be synthesized in parallel on gene chips. For optimal performance in subsequent gene synthesis procedures they should be prepared individually and in larger scales.

Annealing Based Connection of Oligonucleotides

Usually, a set of individually designed oligonucleotides is made on automated solid-phase synthesizers, purified and then connected by specific annealing and standard ligation or polymerase reactions. To improve specificity of oligonucleotide annealing, the synthesis step relies on a set of thermostable DNA ligase and polymerase enzymes. To date, several methods for gene synthesis have been described, such as the ligation of phosphorylated overlapping oligonucleotides, the Fok I method and a modified form of ligase chain reaction for gene synthesis. Additionally, several PCR assembly approaches have been described. They usually employ oligonucleotides of 40-50 nucleotides long that overlap each other. These oligonucleotides are designed to cover most of the sequence of both strands, and the full-length molecule is generated progressively by overlap extension (OE) PCR, thermodynamically balanced inside-out (TBIO) PCR or combined approaches. The most commonly synthesized genes range in size from 600 to 1,200 bp although much longer genes have been made by connecting previously assembled fragments of under 1,000 bp. In this size range it is necessary to test several candidate clones confirming the sequence of the cloned synthetic gene by automated sequencing methods.

Limitations

Moreover, because the assembly of the full-length gene product relies on the efficient and specific alignment of long single stranded oligonucleotides, critical parameters for synthesis success include extended sequence regions comprising secondary structures caused by inverted repeats, extraordinary high or low GC-content, or repetitive structures. Usually these segments of a particular gene can only be synthesized by splitting the procedure into several consecutive steps and a final assembly of shorter sub-sequences, which in turn leads to a significant increase in time and labor needed for its production. The result of a gene synthesis experiment depends strongly on the quality of the oligonucleotides used. For these annealing based gene synthesis protocols, the quality of the product is directly and exponentially dependent on the correctness of the employed oligonucleotides. Alternatively, after performing gene synthesis with oligos of lower quality, more effort must be made in downstream quality assurance during clone analysis, which is usually done by time-consuming standard cloning and sequencing procedures. Another problem associated with all current gene synthesis methods is the high frequency of sequence errors because of the usage of chemically synthesized oligonucleotides. The error frequency increases with longer oligonucleotides, and as a consequence the percentage of correct product decreases dramatically as more oligonucleotides are used. The mutation problem could be solved by shorter oligonucleotides used to assemble the gene. However, all annealing based assembly methods require the primers to be mixed together in one tube. In this case, shorter overlaps do not always allow precise and specific annealing of complementary primers, resulting in the inhibition of full length product formation. Manual design of oligonucleotides is a laborious procedure and does not guarantee the successful synthesis of the desired gene. For optimal performance of almost all annealing based methods, the melting temperatures of the overlapping regions are supposed to be similar for all oligonucleotides. The necessary primer optimization should be performed using specialized oligonucleotide design programs. Several solutions for automated primer design for gene synthesis have been presented so far.

Error Correction Procedures

To overcome problems associated with oligonucleotide quality several elaborate strategies have been developed, employing either separately prepared fishing oligonucleotides, mismatch binding enzymes of the mutS family or specific endonucleases from bacteria or phages. Nevertheless, all these strategies increase time and costs for gene synthesis based on the annealing of chemically synthesized oligonucleotides.

Massively parallel sequencing has also been used as a tool to screen complex oligonucleotide libraries and enable the retrieval of accurate molecules. In one approach, oligonucleotides are sequenced on the 454 pyrosequencing platform and a robotic system images and picks individual beads corresponding to accurate sequence. In another approach, a complex oligonucleotide library is modified with unique flanking tags before massively parallel sequencing. Tag-directed primers then enable the retrieval of molecules with desired sequences by dial-out PCR.

Increasingly, genes are ordered in sets including functionally related genes or multiple sequence variants on a single gene. Virtually all of the therapeutic proteins in development, such as monoclonal antibodies, are optimized by testing many gene variants for improved function or expression.

Applications

Major applications of synthetic genes include synthesis of DNA sequences identified by high throughput sequencing but never cloned into plasmids and the ability to safely obtain genes for vaccine research without the need to grow the full pathogens. Digital manipulation of digital genetic code before synthesis into DNA can be used to optimize protein expression in a particular host, or remove non-functional segments in order to facilitate further replication of the DNA.

Synthesis of DNA allows DNA digital data storage.

DNA Synthesis and Synthetic Biology

The significant drop in cost of gene synthesis in recent years due to increasing competition of companies providing this service has led to the ability to produce entire bacterial plasmids that have never existed in nature. The field of synthetic biology utilizes the technology to produce synthetic biological circuits, which are stretches of DNA manipulated to change gene expression within cells and cause the cell to produce a desired product.

Entire Bacterial Genomes

Synthia and Mycoplasma Laboratorium

A team at the J. Craig Venter Institute published an article in *Science Express*, saying that they had successfully transplanted the natural DNA from a *Mycoplasma mycoides* bacterium into a *Mycoplasma capricolum* cell, creating a bacterium which behaved like a *M. mycoides*.

Craig Venter announced in an interview with UK's *The Guardian* newspaper that the same team had synthesized a modified version of the single chromosome of *Mycoplasma genitalium* using chemicals. The chromosome was modified to eliminate all genes which tests in live bacteria had shown to be unnecessary. The next planned step in this *minimal genome project* is to transplant the synthesized minimal genome into a bacterial cell with its old DNA removed; the resulting bacterium will be called *Mycoplasma laboratorium*. The next day the Canadian bioethics group, ETC Group issued a statement through their representative, Pat Mooney, saying Venter's "creation" was "a chassis on which you could build almost anything". The synthesized genome had not yet been transplanted into a working cell.

On May 21, 2010, *Science* reported that the Venter group had successfully synthesized the genome of the bacterium *Mycoplasma mycoides* from a computer record, and transplanted the synthesized genome into the existing cell of a *Mycoplasma capricolum* bacterium that had had its DNA removed. The "synthetic" bacterium was viable, i.e. capable of replicating billions of times. The team had originally planned to use the *M. genitalium* bacterium they had previously been working with, but switched to *M. mycoides* because the latter bacterium grows much faster, which translated into quicker experiments. Venter describes it as "the first species.... to have its parents be a computer". The transformed bacterium is dubbed "Synthia" by ETC. A Venter spokesperson has declined to confirm any breakthrough at the time of this writing.

Yeast Chromosome

In March 2014, Jef Boeke of the Langone Medical Centre at New York University, published that his team has synthesized one of the *S. cerevisiae* 16 yeast chromosomes, the chromosome III, that he named synIII. The procedure involved replacing the genes in the original chromosome with synthetic versions and the finished human made chromosome was then integrated into a yeast cell. It required designing and creating 273,871 base pairs of DNA - fewer than the 316,667 pairs in the original chromosome.

Unnatural Base Pair (UBP)

DNA sequences have been described which use newly created nucleobases to form a third base pair, in addition to the two base pairs found in nature, A-T (adenine – thymine) and G-C (guanine – cytosine). Multiple research groups have been searching for a third base pair for DNA, including teams led by Steven A. Benner, Philippe Marliere, and Ichiro Hirao. Some new base pairs have been reported.

In 2012, a group of American scientists led by Floyd Romesberg, a chemical biologist at the Scripps Research Institute in San Diego, California, published that his team designed an unnatural base pair (UBP). The two new artificial nucleotides or *Unnatural Base Pair* (UBP) were named d5SICS and dNaM. More technically, these artificial nucleotides bearing hydrophobic nucleobases, feature two fused aromatic rings that form a (d5SICS–dNaM) complex or base pair in DNA. In 2014 the same team from the Scripps Research Institute reported that they synthesized a stretch of circular DNA known as a plasmid containing natural T-A and C-G base pairs along with the best-performing UBP Romesberg's laboratory had designed, and inserted it into cells of the common bacterium *E. coli* that successfully replicated the unnatural base pairs through multiple generations. This is the first known example of a living organism passing along an expanded genetic code to subsequent generations. This was in part achieved by the addition of a supportive algal gene that expresses a nucleotide triphosphate transporter which efficiently imports the triphosphates of both d5SICSTP and dNaMTP into *E. coli* bacteria. Then, the natural bacterial replication pathways use them to accurately replicate the plasmid containing d5SICS–dNaM.

The successful incorporation of a third base pair is a significant breakthrough toward the goal of greatly expanding the number of amino acids which can be encoded by DNA, from the existing 20 amino acids to a theoretically possible 172, thereby expanding the potential for living organisms to produce novel proteins. The artificial strings of DNA do not encode for anything yet, but scientists speculate they could be designed to manufacture new proteins which could have industrial or pharmaceutical uses.

Oligonucleotide Synthesis

Oligonucleotide synthesis is the chemical synthesis of relatively short fragments of nucleic acids with defined chemical structure (sequence). The technique is extremely useful in current laboratory practice because it provides a rapid and inexpensive access to custom-made oligonucleotides of the desired sequence. Whereas enzymes synthesize DNA and RNA only in a 5' to 3' direction, chemical oligonucleotide synthesis does not have this limitation, although it is, most often, carried out in the opposite, 3' to 5' direction. Currently, the process is implemented as solid-phase synthesis using phosphoramidite method and phosphoramidite building blocks derived from protected 2'-deoxynucleosides (dA, dC, dG, and T), ribonucleosides (A, C, G, and U), or chemically modified nucleosides, e.g. LNA or BNA.

To obtain the desired oligonucleotide, the building blocks are sequentially coupled to the growing oligonucleotide chain in the order required by the sequence of the product. The process has been fully automated since the late 1970s. Upon the completion of the chain assembly, the product is released from the solid phase to solution, deprotected, and collected. The occurrence of side reactions sets practical limits for the length of synthetic oligonucleotides (up to about 200 nucleotide residues) because the number of errors accumulates with the length of the oligonucleotide being synthesized. Products are often isolated by high-performance liquid chromatography (HPLC) to obtain the desired oligonucleotides in high purity. Typically, synthetic oligonucleotides are single-stranded DNA or RNA molecules around 15–25 bases in length.

Oligonucleotides find a variety of applications in molecular biology and medicine. They are most commonly used as antisense oligonucleotides, small interfering RNA, primers for DNA sequencing and amplification, probes for detecting complementary DNA or RNA via molecular hybridization, tools for the targeted introduction of mutations and restriction sites, and for the synthesis of artificial genes.

History

The evolution of oligonucleotide synthesis saw four major methods of the formation of internucleosidic linkages and has been reviewed in the literature in great detail.

Early work and Contemporary H-phosphonate Synthesis

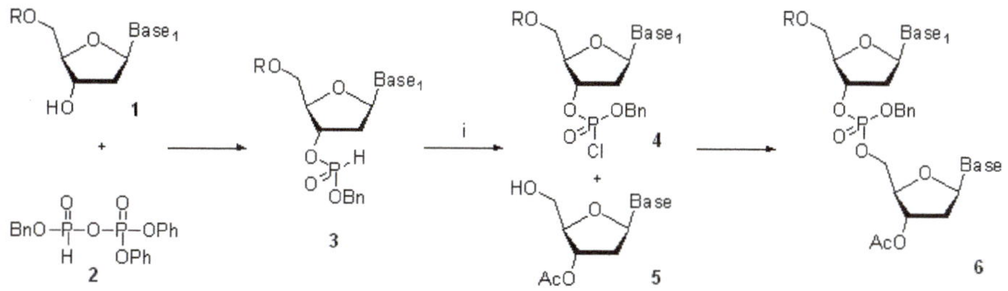

i: *N*-Chlorosuccinimide; Bn = -CH_2Ph

In the early 1950s, Alexander Todd's group pioneered H-phosphonate and phosphate triester methods of oligonucleotide synthesis. The reaction of compounds 1 and 2 to form H-phosphonate diester 3 is an H-phosphonate coupling in solution while that of compounds 4 and 5 to give 6 is a phosphotriester coupling.

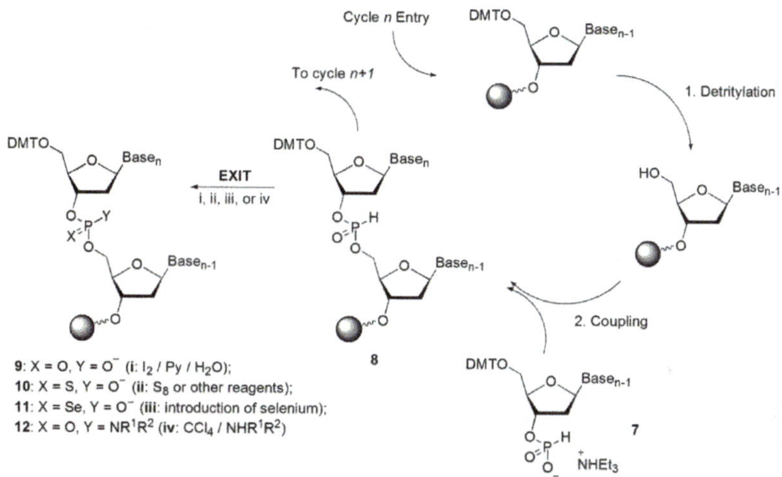

Synthesis of oligonucleotides by the H-Phosphonate Method

Thirty years later, this work inspired, independently, two research groups to adopt the H-phosphonate chemistry to the solid-phase synthesis using nucleoside H-phosphonate monoesters 7 as building blocks and pivaloyl chloride, 2,4,6-triisopropylbenzenesulfonyl chloride (TPS-Cl), and other compounds as activators. The practical implementation of H-phosphonate method resulted in a very short and simple synthetic cycle consisting of only two steps, detritylation and coupling. Oxidation of internucleosidic H-phosphonate diester linkages in 8 to phosphodiester linkages in 9 with a solution of iodine in aqueous pyridine is carried out at the end of the chain assembly rather than as a step in the synthetic cycle. If desired, the oxidation may be carried out under anhydrous conditions. Alternatively, 8 can be converted to phosphorothioate 10 or phosphoroselenoate 11 (X = Se), or oxidized by CCl_4 in the presence of primary or secondary amines to phosphoramidate analogs 12. The method is very convenient in that various types of phosphate modifications (phosphate/phosphorothioate/phosphoramidate) may be introduced to the same oligonucleotide for modulation of its properties.

Most often, H-phosphonate building blocks are protected at the 5'-hydroxy group and at the amino group of nucleic bases A, C, and G in the same manner as phosphoramidite building blocks. However, protection at the amino group is not mandatory.

Phosphodiester Synthesis

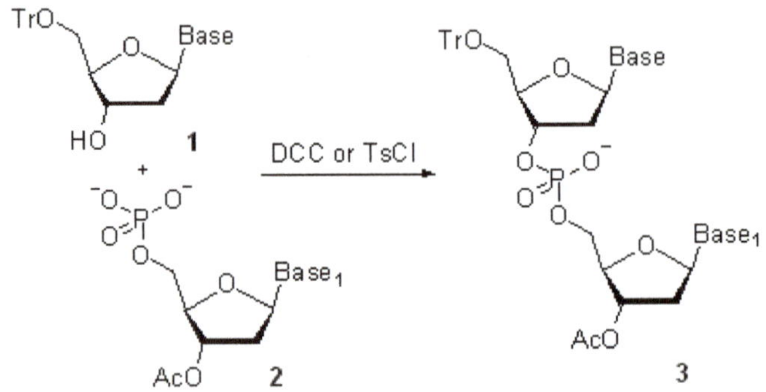

Oligonucleotide coupling by phosphodiester method; Tr = $-CPh_3$

In the 1950s, Har Gobind Khorana and co-workers developed a phosphodiester method where 3'-*O*-acetylnucleoside-5'-*O*-phosphate 2 was activated with *N,N'*-dicyclohexylcarbodiimide (DCC) or 4-toluenesulfonyl chloride (Ts-Cl). The activated species were reacted with a 5'-*O*-protected nucleoside 1 to give a protected dinucleoside monophosphate 3. Upon the removal of 3'-*O*-acetyl group using base-catalyzed hydrolysis, further chain elongation was carried out. Following this methodology, sets of tri- and tetradeoxyribonucleotides were synthesized and were enzymatically converted to longer oligonucleotides, which allowed elucidation of the genetic code. The major limitation of the phosphodiester method consisted in the formation of pyrophosphate oligomers and oligonucleotides branched at the internucleosidic phosphate. The method seems to be a step back from the more selective chemistry described earlier; however, at that time, most phosphate-protecting groups available now had not yet been introduced. The lack of the convenient protection strategy necessitated taking a retreat to a slower and less selective chemistry to achieve the ultimate goal of the study.

Phosphotriester Synthesis

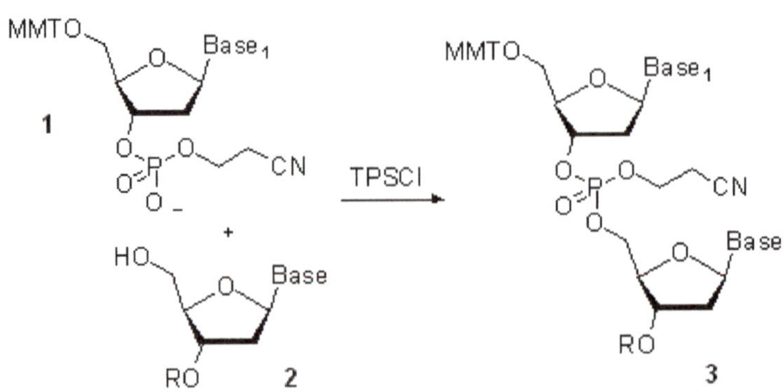

Oligonucleotide coupling by phosphotriester method; MMT = $-CPh_2(4\text{-}MeOC_6H_4)$.

In the 1960s, groups led by R. Letsinger and C. Reese developed a phosphotriester approach. The defining difference from the phosphodiester approach was the protection of the phosphate moiety in the building block 1 and in the product 3 with 2-cyanoethyl group. This precluded the formation of oligonucleotides branched at the internucleosidic phosphate. The higher selectivity of the method allowed the use of more efficient coupling agents and catalysts, which dramatically reduced the length of the synthesis. The method, initially developed for the solution-phase synthesis, was also implemented on low-cross-linked "popcorn" polystyrene, and later on controlled pore glass, which initiated a massive research effort in solid-phase synthesis of oligonucleotides and eventually led to the automation of the oligonucleotide chain assembly.

Phosphite Triester Synthesis

In the 1970s, substantially more reactive P(III) derivatives of nucleosides, 3'-*O*-chlorophosphites, were successfully used for the formation of internucleosidic linkages. This led to the discovery of the phosphite triester methodology. The group led by M. Caruthers took the advantage of less aggressive and more selective 1*H*-tetrazolidophosphites and implemented the method on solid phase. Very shortly after, the workers from the same group further improved the method by using more stable nucleoside phosphoramidites as building blocks. The use of 2-cyanoethyl phosphite-protecting group in place of a less user-friendly methyl group led to the nucleoside phosphoramidites currently used in oligonucleotide synthesis. Many later improvements to the manufacturing of building blocks, oligonucleotide synthesizers, and synthetic protocols made the phosphoramidite chemistry a very reliable and expedient method of choice for the preparation of synthetic oligonucleotides.

Synthesis by the Phosphoramidite Method

Building Blocks

Nucleoside Phosphoramidites

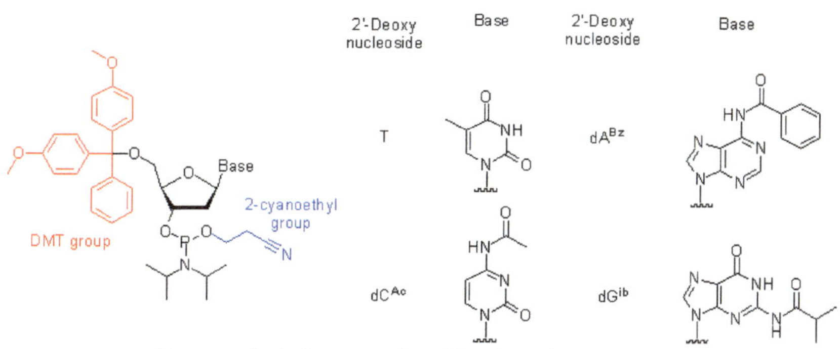

Protected 2'-deoxynucleoside phosphoramidites.

As mentioned above, the naturally occurring nucleotides (nucleoside-3'- or 5'-phosphates) and their phosphodiester analogs are insufficiently reactive to afford an expeditious synthetic preparation of oligonucleotides in high yields. The selectivity and the rate of the formation of internucleosidic linkages is dramatically improved by using 3'-*O*-(*N,N*-diisopropyl phosphoramidite) derivatives of nucleosides (nucleoside phosphoramidites) that serve as building blocks in phosphite triester methodology. To prevent undesired side reactions, all other functional groups present in nucleosides have to be rendered unreactive (protected) by attaching protecting groups. Upon the completion of the oligonucleotide chain assembly, all the protecting groups are removed to yield the desired oligonucleotides. Below, the protecting groups currently used in commercially available and most common nucleoside phosphoramidite building blocks are briefly reviewed:

- The 5'-hydroxyl group is protected by an acid-labile DMT (4,4'-dimethoxytrityl) group.
- Thymine and uracil, nucleic bases of thymidine and uridine, respectively, do not have exocyclic amino groups and hence do not require any protection.
- Although the nucleic base of guanosine and 2'-deoxyguanosine does have an exocyclic amino group, its basicity is low to an extent that it does not react with phosphoramidites under the conditions of the coupling reaction. However, a phosphoramidite derived from the N2-unprotected 5'-*O*-DMT-2'-deoxyguanosine is poorly soluble in acetonitrile, the solvent commonly used in oligonucleotide synthesis. In contrast, the N2-protected versions of the same compound dissolve in acetonitrile well and hence are widely used. Nucleic bases adenine and cytosine bear the exocyclic amino groups reactive with the activated phosphoramidites under the conditions of the coupling reaction. By the use of additional steps in the synthetic cycle or alternative coupling agents and solvent systems, the oligonucleotide chain assembly may be carried out using dA and dC phosphoramidites with unprotected amino groups. However, these approaches currently remain in the research stage. In routine oligonucleotide synthesis, exocyclic amino groups in nucleosides are kept permanently protected over the entire length of the oligonucleotide chain assembly.

The protection of the exocyclic amino groups has to be orthogonal to that of the 5'-hydroxy group because the latter is removed at the end of each synthetic cycle. The simplest to implement, and hence the most widely used, strategy is including a base-labile protection group on the exocyclic amino groups. Most often, two protection schemes are used.

- In the first, the standard and more robust scheme, Bz (benzoyl) protection is used for A, dA, C, and dC, while G and dG are protected with isobutyryl group. More recently, Ac (acetyl) group is used to protect C and dC as shown in figure.
- In the second, mild protection scheme, A and dA are protected with isobutyryl or phenoxyacetyl groups (PAC). C and dC bear acetyl protection, and G and dG are protected with 4-isopropylphenoxyacetyl (*i*Pr-PAC) or dimethyl-

formamidino (dmf) groups. Mild protecting groups are removed more readily than the standard protecting groups. However, the phosphoramidites bearing these groups are less stable when stored in solution.

- The phosphite group is protected by a base-labile 2-cyanoethyl group. Once a phosphoramidite has been coupled to the solid support-bound oligonucleotide and the phosphite moieties have been converted to the P(V) species, the presence of the phosphate protection is not mandatory for the successful conducting of further coupling reactions.

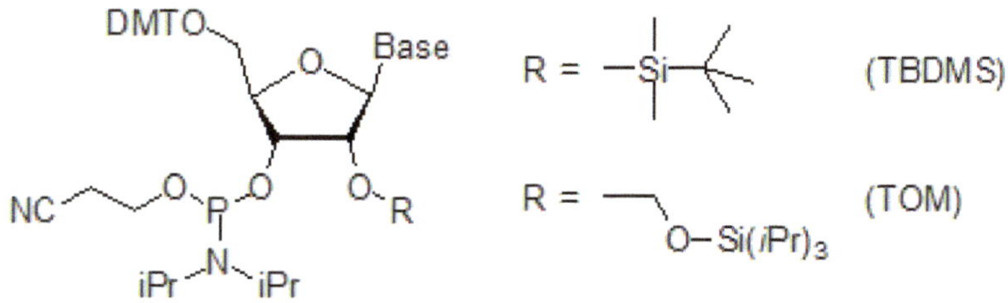

2'-*O*-protected ribonucleoside phosphoramidites.

- In RNA synthesis, the 2'-hydroxy group is protected with TBDMS (*t*-butyldimethylsilyl) group. or with TOM (tri-*iso*-propylsilyloxymethyl) group, both being removable by treatment with fluoride ion.

- The phosphite moiety also bears a diisopropylamino (iPr_2N) group reactive under acidic conditions. Upon activation, the diisopropylamino group leaves to be substituted by the 5'-hydroxy group of the support-bound oligonucleotide.

Non-nucleoside Phosphoramidites

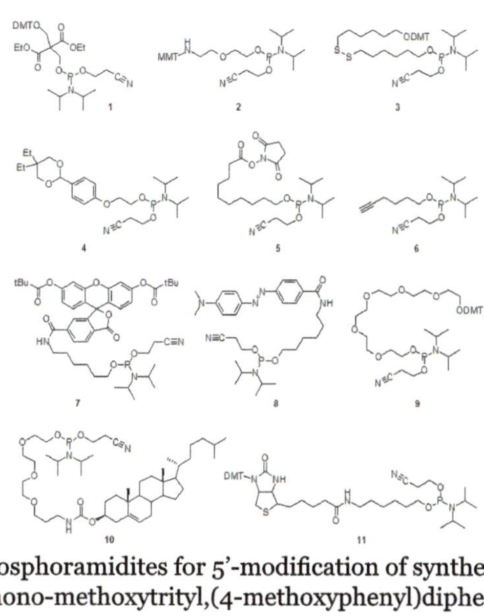

Non-nucleoside phosphoramidites for 5'-modification of synthetic oligonucleotides. MMT = mono-methoxytrityl,(4-methoxyphenyl)diphenylmethyl.

Non-nucleoside phosphoramidites are the phosphoramidite reagents designed to introduce various functionalities at the termini of synthetic oligonucleotides or between nucleotide residues in the middle of the sequence. In order to be introduced inside the sequence, a non-nucleosidic modifier has to possess at least two hydroxy groups, one of which is often protected with the DMT group while the other bears the reactive phosphoramidite moiety.

Non-nucleosidic phosphoramidites are used to introduce desired groups that are not available in natural nucleosides or that can be introduced more readily using simpler chemical designs. A very short selection of commercial phosphoramidite reagents is shown in scheme for the demonstration of the available structural and functional diversity. These reagents serve for the attachment of 5'-terminal phosphate (1), NH_2 (2), SH (3), aldehydo (4), and carboxylic groups (5), CC triple bonds (6), non-radioactive labels and quenchers (exemplified by 6-FAM amidite 7 for the attachment of fluorescein and dabcyl amidite 8, respectively), hydrophilic and hydrophobic modifiers (exemplified by hexaethyleneglycol amidite 9 and cholesterol amidite 10, respectively), and biotin amidite 11.

Synthetic Cycle

Oligonucleotide synthesis is carried out by a stepwise addition of nucleotide residues to the 5'-terminus of the growing chain until the desired sequence is assembled. Each addition is referred to as a synthetic cycle and consists of four chemical reactions:

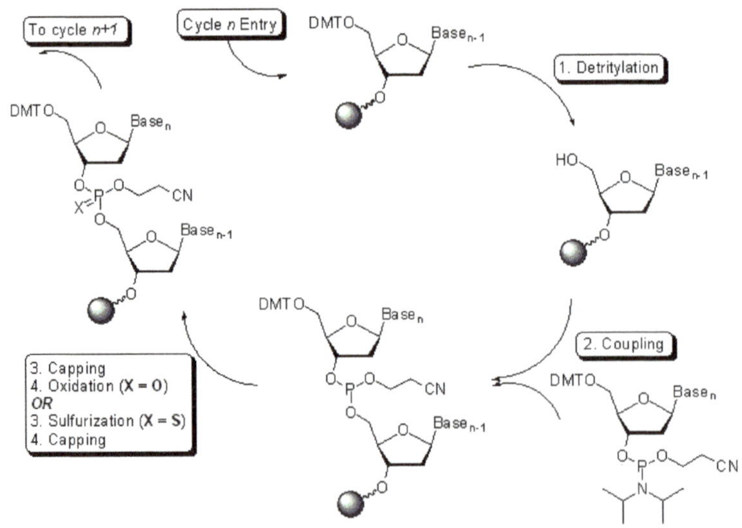

Synthetic cycle for preparation of oligonucleotides by phosphoramidite method.

Step 1: De-blocking (Detritylation)

The DMT group is removed with a solution of an acid, such as 2% trichloroacetic acid (TCA) or 3% dichloroacetic acid (DCA), in an inert solvent (dichloromethane or toluene). The orange-colored DMT cation formed is washed out; the step results in the solid support-bound oligonucleotide precursor bearing a free 5'-terminal hydroxyl group. It is worth remembering that conducting detritylation for an extended time or with stronger than recommended solutions of acids leads to depurination of solid support-bound oligonucleotide and thus reduces the yield of the desired full-length product.

Step 2: Coupling

A 0.02–0.2 M solution of nucleoside phosphoramidite (or a mixture of several phosphoramidites) in acetonitrile is activated by a 0.2–0.7 M solution of an acidic azole catalyst, 1*H*-tetrazole, 5-ethylthio-1H-tetrazole, 2-benzylthiotetrazole, 4,5-dicyanoimidazole, or a number of similar compounds. A more extensive information on the use of various coupling agents in oligonucleotide synthesis can be found in a recent review. The mixing is usually very brief and occurs in fluid lines of oligonucleotide synthesizers while the components are being delivered to the reactors containing solid support. The activated phosphoramidite in 1.5 – 20-fold excess over the support-bound material is then brought in contact with the starting solid support (first coupling) or a support-bound oligonucleotide precursor (following couplings) whose 5'-hydroxy group reacts with the activated phosphoramidite moiety of the incoming nucleoside phosphoramidite to form a phosphite triester linkage. The coupling of 2'-deoxynucleoside phosphoramidites is very rapid and requires, on small scale, about 20 s for its completion. In contrast, sterically hindered 2'-*O*-protected ribonucleoside phosphoramidites require 5-15 min to be coupled in high yields. The reaction is also highly sensitive to the presence of water, particularly when dilute solutions of phosphoramidites are used, and is commonly carried out in anhydrous acetonitrile. Generally, the larger the scale of the synthesis, the lower the excess and the higher the concentration of the phosphoramidites is used. In contrast, the concentration of the activator is primarily determined by its solubility in acetonitrile and is irrespective of the scale of the synthesis. Upon the completion of the coupling, any unbound reagents and by-products are removed by washing.

Step 3: Capping

The capping step is performed by treating the solid support-bound material with a mixture of acetic anhydride and 1-methylimidazole or, less often, DMAP as catalysts and, in the phosphoramidite method, serves two purposes.

- After the completion of the coupling reaction, a small percentage of the solid support-bound 5'-OH groups (0.1 to 1%) remains unreacted and needs to be permanently blocked from further chain elongation to prevent the formation of oligonucleotides with an internal base deletion commonly referred to as (n-1) shortmers. The unreacted 5'-hydroxy groups are, to a large extent, acetylated by the capping mixture.

- It has also been reported that phosphoramidites activated with 1*H*-tetrazole react, to a small extent, with the O^6 position of guanosine. Upon oxidation with I_2 /water, this side product, possibly via O^6-N7 migration, undergoes depurination. The apurinic sites thus formed are readily cleaved in the course of the final deprotection of the oligonucleotide under the basic conditions to give two shorter oligonucleotides thus reducing the yield of the full-length product. The O^6 modifications are rapidly removed by treatment with the capping reagent as long as the capping step is performed *prior* to oxidation with I_2/water.

- The synthesis of oligonucleotide phosphorothioates does not involve the oxidation with I_2/water, and, respectively, does not suffer from the side reaction described above. On the other hand, if the capping step is performed prior to sulfurization, the solid support may contain the residual acetic anhydride and N-methylimidazole left after the capping step. The capping mixture interferes with the sulfur transfer reaction, which results in the extensive formation of the phosphate triester internucleosidic linkages in place of the desired PS triesters. Therefore, for the synthesis of OPS, it is advisable to conduct the sulfurization step *prior* to the capping step.

Step 4: Oxidation

The newly formed tricoordinated phosphite triester linkage is not natural and is of limited stability under the conditions of oligonucleotide synthesis. The treatment of the support-bound material with iodine and water in the presence of a weak base (pyridine, lutidine, or collidine) oxidizes the phosphite triester into a tetracoordinated phosphate triester, a protected precursor of the naturally occurring phosphate diester internucleosidic linkage. Oxidation may be carried out under anhydrous conditions using tert-Butyl hydroperoxide or, more efficiently, (1S)-(+)-(10-camphorsulfonyl)-oxaziridine (CSO). The step of oxidation is substituted with a sulfurization step to obtain oligonucleotide phosphorothioates. In the latter case, the sulfurization step is best carried out prior to capping.

Solid Supports

In solid-phase synthesis, an oligonucleotide being assembled is covalently bound, via its 3'-terminal hydroxy group, to a solid support material and remains attached to it over the entire course of the chain assembly. The solid support is contained in columns whose dimensions depend on the scale of synthesis and may vary between 0.05 mL and several liters. The overwhelming majority of oligonucleotides are synthesized on small scale ranging from 40 nmol to 1 μmol. More recently, high-throughput oligonucleotide synthesis where the solid support is contained in the wells of multi-well plates (most often, 96 or 384 wells per plate) became a method of choice for parallel synthesis of oligonucleotides on small scale. At the end of the chain assembly, the oligonucleotide is released from the solid support and is eluted from the column or the well.

Solid Support Material

In contrast to organic solid-phase synthesis and peptide synthesis, the synthesis of oligonucleotides proceeds best on non-swellable or low-swellable solid supports. The two most often used solid-phase materials are controlled pore glass (CPG) and macroporous polystyrene (MPPS).

- CPG is commonly defined by its pore size. In oligonucleotide chemistry, pore sizes of 500, 1000, 1500, 2000, and 3000 Å are used to allow the preparation of about 50, 80, 100, 150, and 200-mer oligonucleotides, respectively. To make native CPG suitable for further processing, the surface of the material is treated with (3-aminopropyl)triethoxysilane to give aminopropyl CPG. The aminopropyl arm may be further extended to result in long chain aminoalkyl (LCAA) CPG. The amino group is then used as an anchoring point for linkers suitable for oligonucleotide synthesis.
- MPPS suitable for oligonucleotide synthesis is a low-swellable, highly cross-linked polystyrene obtained by polymerization of divinylbenzene (min 60%), styrene, and 4-chloromethylstyrene in the presence of a porogeneous agent. The macroporous chloromethyl MPPS obtained is converted to aminomethyl MPPS.

Linker Chemistry

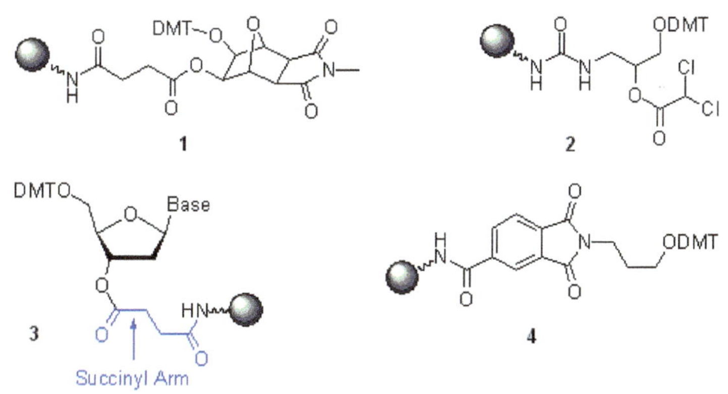

Commercial solid supports for oligonucleotide synthesis.

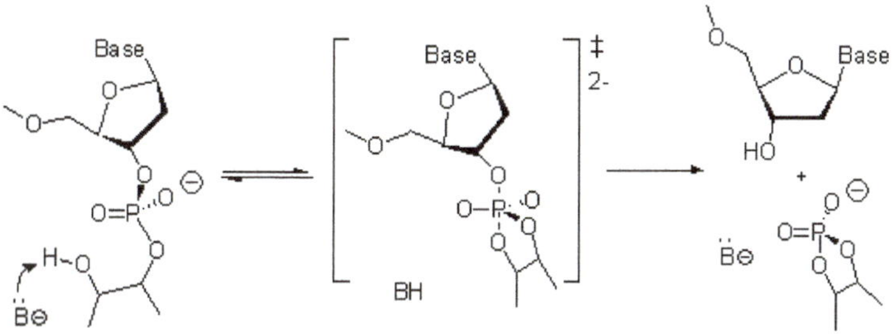

Mechanism of 3'-dephosphorylation of oligonucleotides assembled on universal solid supports.

To make the solid support material suitable for oligonucleotide synthesis, non-nucleosidic linkers or nucleoside succinates are covalently attached to the reactive amino groups in aminopropyl CPG, LCAA CPG, or aminomethyl MPPS. The remaining unreacted amino groups are capped with acetic anhydride. Typically, three conceptually different groups of solid supports are used.

- *Universal supports.* In a more recent, more convenient, and more widely used method, the synthesis starts with the universal support where a non-nucleosidic linker is attached to the solid support material (compounds 1 and 2). A phosphoramidite respective to the 3'-terminal nucleoside residue is coupled to the universal solid support in the first synthetic cycle of oligonucleotide chain assembly using the standard protocols. The chain assembly is then continued until the completion, after which the solid support-bound oligonucleotide is deprotected. The characteristic feature of the universal solid supports is that the release of the oligonucleotides occurs by the hydrolytic cleavage of a P-O bond that attaches the 3'-*O* of the 3'-terminal nucleotide residue to the universal linker. The critical advantage of this approach is that the same solid support is used irrespectively of the sequence of the oligonucleotide to be synthesized. For the complete removal of the linker and the 3'-terminal phosphate from the assembled oligonucleotide, the solid support 1 and several similar solid supports require gaseous ammonia, aqueous ammonium hydroxide, aqueous methylamine, or their mixture and are commercially available. The solid support 2 requires a solution of ammonia in anhydrous methanol and is also commercially available.
- *Nucleosidic solid supports.* In a historically first and still popular approach, the 3'-hydroxy group of the 3'-terminal nucleoside residue is attached to the solid support via, most often, 3'-*O*-succinyl arm as in compound 3. The oligonucleotide chain assembly starts with the coupling of a phosphoramidite building block respective to the nucleotide residue second from the 3'-terminus. The 3'-terminal hydroxy group in oligonucleotides synthesized on nucleosidic solid supports is deprotected under the conditions somewhat milder than those applicable for universal solid supports. However, the fact that a nucleosidic solid support has to be selected in a sequence-specific manner reduces the throughput of the entire synthetic process and increases the likelihood of human error.
- *Special solid supports* are used for the attachment of desired functional or reporter groups at the 3'-terminus of synthetic oligonucleotides. For example, the commercial solid support 4 allows the preparation of oligonucleotides bearing 3'-terminal 3-aminopropyl linker. Similarly to non-nucleosidic phosphoramidites, many other special solid supports designed for the attachment of reactive functional groups, non-radioactive reporter groups, and terminal modifiers (*e.c.* cholesterol or other hydrophobic tethers) and suited for various applications are commercially available. A more detailed information on various solid supports for oligonucleotide synthesis can be found in a recent review.

Oligonucleotide Phosphorothioates and their Synthesis

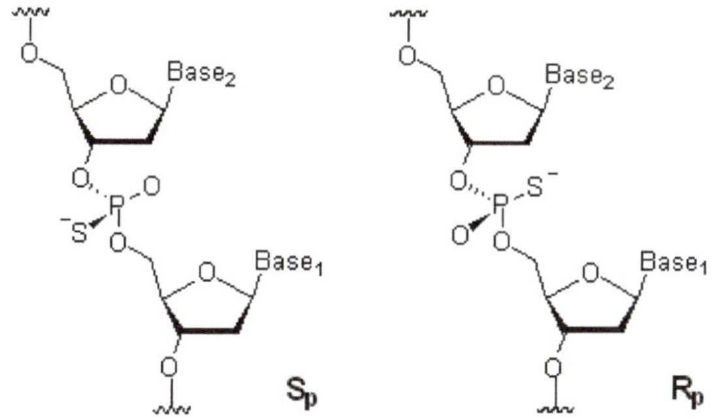

S_p and R_p-diastereomeric internucleosidic phosphorothioate linkages.

Oligonucleotide phosphorothioates (OPS) are modified oligonucleotides where one of the oxygen atoms in the phosphate moiety is replaced by sulfur. Only the phosphorothioates having sulfur at a non-bridging position as shown in figure are widely used and are available commercially. The replacement of the non-bridging oxygen with sulfur creates a new center of chirality at phosphorus. In a simple case of a dinucleotide, this results in the formation of a diastereomeric pair of S_p- and R_p-dinucleoside monophosphorothioates whose structures are shown in figure. In an n-mer oligonucleotide where all $(n - 1)$ internucleosidic linkages are phosphorothioate linkages, the number of diastereomers m is calculated as $m = 2^{(n-1)}$. Being non-natural analogs of nucleic acids, OPS are substantially more stable towards hydrolysis by nucleases, the class of enzymes that destroy nucleic acids by breaking the bridging P-O bond of the phosphodiester moiety. This property determines the use of OPS as antisense oligonucleotides in *in vitro* and *in vivo* applications where the extensive exposure to nucleases is inevitable. Similarly, to improve the stability of siRNA, at least one phosphorothioate linkage is often introduced at the 3'-terminus of both sense and antisense strands. In chirally pure OPS, all-Sp diastereomers are more stable to enzymatic degradation than their all-Rp analogs. However, the preparation of chirally pure OPS remains a synthetic challenge. In laboratory practice, mixtures of diastereomers of OPS are commonly used.

Synthesis of OPS is very similar to that of natural oligonucleotides. The difference is that the oxidation step is replaced by sulfur transfer reaction (sulfurization) and that the capping step is performed after the sulfurization. Of many reported reagents capable of the efficient sulfur transfer, only three are commercially available:

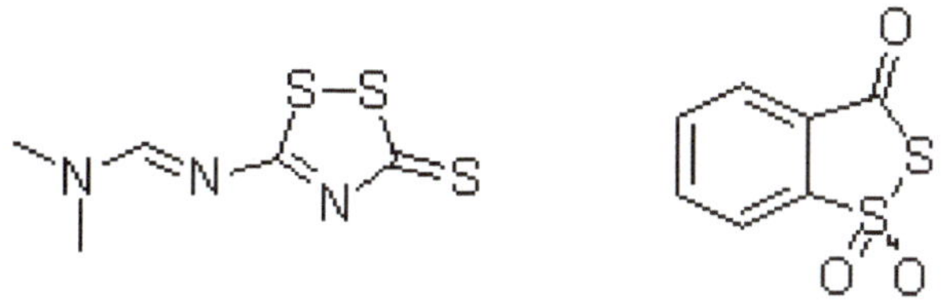

3 4

Commercial sulfur transfer agents for oligonucleotide synthesis.

- 3-(Dimethylaminomethylidene)amino-3H-1,2,4-dithiazole-3-thione, DDTT (3) provides rapid kinetics of sulfurization and high stability in solution. The reagent is available from several sources.
- 3*H*-1,2-benzodithiol-3-one 1,1-dioxide (4) also known as Beaucage reagent displays a better solubility in acetonitrile and short reaction times. However, the reagent is of limited stability in solution and is less efficient in sulfurizing RNA linkages.
- *N,N,N'N'*-Tetraethylthiuram disulfide (TETD) is soluble in acetonitrile and is commercially available. However, the sulfurization reaction of an internucleosidic DNA linkage with TETD requires 15 min, which is more than 10 times as slow as that with compounds 3 and 4.

Automation

In the past, oligonucleotide synthesis was carried out manually in solution or on solid phase. The solid phase synthesis was implemented using, as containers for the solid phase, miniature glass columns similar in their shape to low-pressure chromatography columns or syringes equipped with porous filters. Currently, solid-phase oligonucleotide synthesis is carried out automatically using computer-controlled instruments (oligonucleotide synthesizers) and is technically implemented in column, multi-well plate, and array formats. The column format is best suited for research and large scale applications where a high-throughput is not required. Multi-well plate format is designed specifically for high-throughput synthesis on small scale to satisfy the growing demand of industry and academia for synthetic oligonucleotides. A number of oligonucleotide synthesizers for small scale synthesis and medium to large scale synthesis are available commercially.

First Commercially Available Oligonucleotide Synthesizers

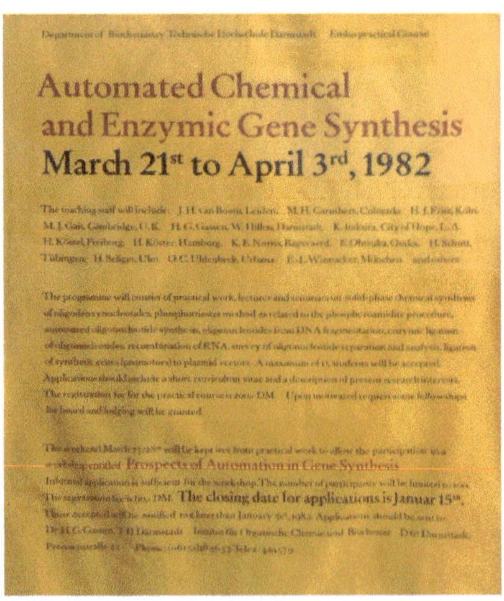

In March 1982 a practical course was hosted by the Department of Biochemistry, Technische Hochschule Darmstadt, Germany. M.H. Caruthers, M.J. Gait, H.G. Gassen, H.Koster, K. Itakura, and C. Birr among others attended. The program comprised practical work, lectures, and seminars on solid-phase chemical synthesis of oligonucleotides. A select group of 15 students attended and had an unprecedented opportunity to be instructed by the esteemed teaching staff.

Along with manual exercises, several prominent automation companies attended the course. Biosearch of Novato, CA, Genetic Design of Watertown, MA, were two of several companies to demonstrate automated synthesizers at the course. Biosearch presented their new SAM I synthesizer. The Genetic Design had developed their synthesizer from the design of its sister companies (Sequemat) solid phase peptide sequencer. The Genetic Design arranged with Dr Christian Birr (Max-Planck-Institute for Medical Research) a week before the event to convert his solid phase sequencer into the semi-automated synthesizer. The team led by Dr Alex Bonner and Rick Neves converted the unit and transported it to Darmstadt for the event and installed into the Biochemistry lab at the Technische Hochschule. As the system was semi-automatic, the user injected the next base to be added to the growing sequence during each cycle. The system worked well and produced a series of test tubes filled with bright red trityl color indicating complete coupling at each step. This system was later fully automated by inclusion of an auto injector and was designated the Model 25A.

History of Mid to Large Scale Oligonucleotide Synthesis

Large scale oligonucleotide synthesizers were often developed by augmenting the capabilities of a preexisting instrument platform. One of the first mid scale synthesizers appeared in the late 1980s, manufactured by the Biosearch company in Novato, CA (The 8800). This platform was originally designed as a peptide synthesizer and made use of a fluidized bed reactor essential for accommodating the swelling characteristics of polystyrene supports used in the Merrifield methodology. Oligonucleotide synthesis involved the use of CPG (controlled pore glass) which is a rigid support and is more suited for column reactors as described above. The scale of the 8800 was limited to the flow rate required to fluidize the support. Some novel reactor designs as well as higher than normal pressures enabled the 8800 to achieve scales that would prepare 1 mmole of oligonucleotide. In the mid 1990s several companies developed platforms that were based on semi-preparative and preparative liquid chromatographs. These systems were well suited for a column reactor approach. In most cases all that was required was to augment the number of fluids that could be delivered to the column. Oligo synthesis requires a minimum of 10 and liquid chromatographs usually accommodate 4. This was an easy design task and some semi-automatic strategies worked without any modifications to the preexisting LC equipment. PerSeptive Biosystems as well as Pharmacia (GE) were two of several companies that developed synthesizers out of liquid chromatographs. Genomic Technologies,

Inc. was one of the few companies to develop a large scale oligonucleotide synthesizer that was, from the ground up, an oligonucleotide synthesizer. The initial platform called the VLSS for very large scale synthesizer utilized large Pharmacia liquid chromatograph columns as reactors and could synthesize up to 75 millimoles of material. Many oligonucleotide synthesis factories designed and manufactured their own custom platforms and little is known due to the designs being proprietary. The VLSS design continued to be refined and is continued in the QMaster synthesizer which is a scaled down platform providing milligram to gram amounts of synthetic oligonucleotide.

The current practices of synthesis of chemically modified oligonucleotides on large scale have been recently reviewed.

Synthesis of Oligonucleotide Microarrays

One may visualize an oligonucleotide microarray as a miniature multi-well plate where physical dividers between the wells (plastic walls) are intentionally removed. With respect to the chemistry, synthesis of oligonucleotide microarrays is different from the conventional oligonucleotide synthesis in two respects:

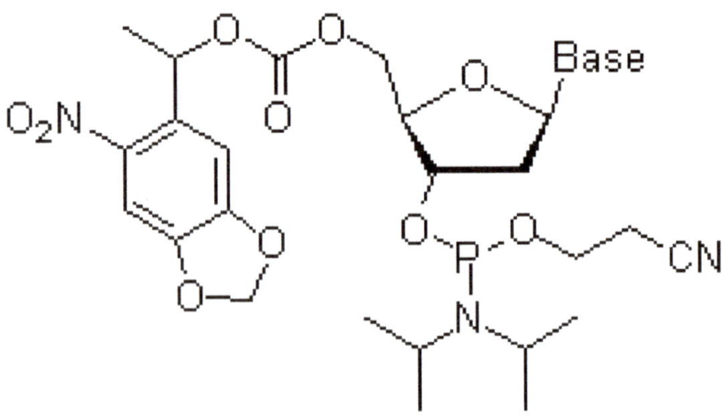

5'-*O*-MeNPOC-protected nucleoside phosphoramidite.

- Oligonucleotides remain permanently attached to the solid phase, which requires the use of linkers that are stable under the conditions of the final deprotection procedure.
- The absence of physical dividers between the sites occupied by individual oligonucleotides, a very limited space on the surface of the microarray (one oligonucleotide sequence occupies a square 25×25 μm) and the requirement of high fidelity of oligonucleotide synthesis dictate the use of site-selective 5'-deprotection techniques. In one approach, the removal of the 5'-*O*-DMT group is effected by electrochemical generation of the acid at the required site(s). Another approach uses 5'-*O*-(α-methyl-6-nitropiperonyloxycarbonyl) (MeNPOC) protecting group, which can be removed by irradiation with UV light of 365 nm wavelength.

Post-synthetic Processing

After the completion of the chain assembly, the solid support-bound oligonucleotide is fully protected:

- The 5'-terminal 5'-hydroxy group is protected with DMT group;
- The internucleosidic phosphate or phosphorothioate moieties are protected with 2-cyanoethyl groups;
- The exocyclic amino groups in all nucleic bases except for T and U are protected with acyl protecting groups.

To furnish a functional oligonucleotide, all the protecting groups have to be removed. The N-acyl base protection and the 2-cyanoethyl phosphate protection may be, and is often removed simultaneously by treatment with inorganic bases or amines. However, the applicability of this method is limited by the fact that the cleavage of 2-cyanoethyl phosphate protection gives rise to acrylonitrile as a side product. Under the strong basic conditions required for the removal of N-acyl protection, acrylonitrile is capable of alkylation of nucleic bases, primarily, at the N3-position of thymine and uracil residues to give the respective N3-(2-cyanoethyl) adducts via Michael reaction. The formation of these side products may be avoided by treating the solid support-bound oligonucleotides with solutions of bases in an organic solvent, for instance, with 50% triethylamine in acetonitrile or 10% diethylamine in acetonitrile. This treatment is strongly recommended for medium- and large scale preparations and is optional for syntheses on small scale where the concentration of acrylonitrile generated in the deprotection mixture is low.

Regardless of whether the phosphate protecting groups were removed first, the solid support-bound oligonucleotides are deprotected using one of the two general approaches.

- (1) Most often, 5'-DMT group is removed at the end of the oligonucleotide chain assembly. The oligonucleotides are then released from the solid phase and deprotected (base and phosphate) by treatment with aqueous ammonium hydroxide, aqueous methylamine, their mixtures, gaseous ammonia or methylamine or, less commonly, solutions of other primary amines or alkalies at ambient or elevated temperature. This removes all remaining protection groups from 2'-deoxyoligonucleotides, resulting in a reaction mixture containing the desired product. If the oligonucleotide contains any 2'-*O*-protected ribonucleotide residues, the deprotection protocol includes the second step where the 2'-*O*-protecting silyl groups are removed by treatment with fluoride ion by various methods. The fully deprotected product is used as is, or the desired oligonucleotide can be purified by a number of methods. Most commonly, the crude product is desalted using ethanol precipitation, size exclusion chromatography, or reverse-phase HPLC. To eliminate unwanted truncation products, the oligonucleotides can be purified via polyacrylamide gel electrophoresis or anion-exchange HPLC followed by desalting.

- (2) The second approach is only used when the intended method of purification is reverse-phase HPLC. In this case, the 5'-terminal DMT group that serves as a hydrophobic handle for purification is kept on at the end of the synthesis. The oligonucleotide is deprotected under basic conditions as described above and, upon evaporation, is purified by reverse-phase HPLC. The collected material is then detritylated under aqueous acidic conditions. On small scale (less than 0.01–0.02 mmol), the treatment with 80% aqueous acetic acid for 15–30 min at room temperature is often used followed by evaporation of the reaction mixture to dryness *in vacuo*. Finally, the product is desalted as described above.
- For some applications, additional reporter groups may be attached to an oligonucleotide using a variety of post-synthetic procedures.

Characterization

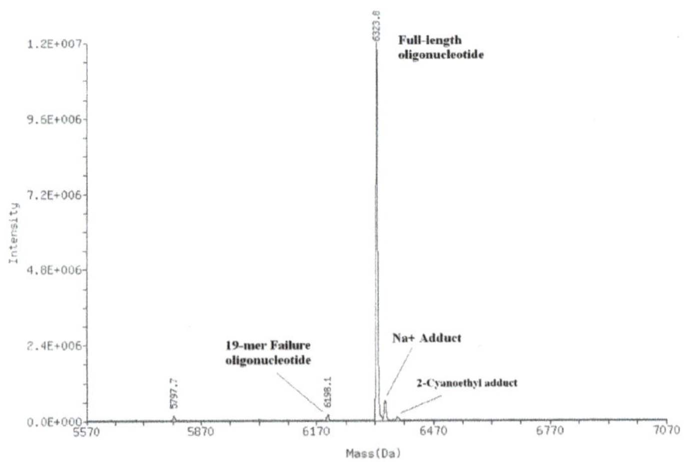

Deconvoluted ES MS of crude oligonucleotide 5'-DMT-T_{20} (calculated mass 6324.26 Da).

As with any other organic compound, it is prudent to characterize synthetic oligonucleotides upon their preparation. In more complex cases (research and large scale syntheses) oligonucleotides are characterized after their deprotection and after purification. Although the ultimate approach to the characterization is sequencing, a relatively inexpensive and routine procedure, the considerations of the cost reduction preclude its use in routine manufacturing of oligonucleotides. In day-by-day practice, it is sufficient to obtain the molecular mass of an oligonucleotide by recording its mass spectrum. Two methods are currently widely used for characterization of oligonucleotides: electrospray mass spectrometry (ES MS) and matrix-assisted laser desorption/ionization time-of-flight mass spectrometry (MALDI-TOF). To obtain informative spectra, it is very important to exchange all metal ions that might be present in the sample for ammonium or trialkylammonium [*e.c.* triethylammonium, $(C_2H_5)_3NH^+$] ions prior to submitting a sample to the analysis by either of the methods.

- In ES MS spectrum, a given oligonucleotide generates a set of ions that correspond to different ionization states of the compound. Thus, the oligonucleotide with molecular mass M generates ions with masses $(M - nH)/n$ where M is the molecular mass of the oligonucleotide in the form of a free acid (all negative charges of internucleosidic phosphodiester groups are neutralized with H^+), n is the ionization state, and H is the atomic mass of hydrogen (1 Da). Most useful for characterization are the ions with n ranging from 2 to 5. Software supplied with the more recently manufactured instruments is capable of performing a deconvolution procedure that is, it finds peaks of ions that belong to the same set and derives the molecular mass of the oligonucleotide.
- To obtain more detailed information on the impurity profile of oligonucleotides, liquid chromatography-mass spectrometry (LC-MS or HPLC-MS) or capillary electrophoresis mass spectrometry (CEMS) are used.

Cre-lox System

This system allows the genetic manipulation of target cells to control its gene expression, delete specific DNA sequences, or modify the genomic content. The *Cre* recombinase is a site-specific integrase isolated from bacteriophage P1. It catalyzes the recombination of DNA between specific *loxP* sites in DNA. Generally, this system is created after generating two strains, one expressing *Cre* recombinase and the other having *loxP* site flanked with the gene of interest. Both the strains are crossed in order to allow independent recombination and their outcome is determined by the location and orientation of *loxP* site containing gene of interest. If the *loxP* sites are oriented in opposite directions, *Cre* recombinase mediates the inversion of the gene of interest. However, *Cre* recombinase mediates a translocation event if the *loxP* sites are located on different chromosomes and deletion event if the *loxP* sites are oriented in the same direction on a chromosome segment.

Some terminology:

Pluripotent embryonic stem cells are undifferentiated early embryonic cells derived from the inner cell mass of mouse blastocysts.

Nuclear localization sequences (NLS) are important for directing a protein to nucleus. Generally NLS contains PK3RKV amino acid residues in the protein. The proteins are directed to the nucleus with the help of nuclear pore complex and with the combined effect of RAN-GTPase and IMPORTIN molecules.

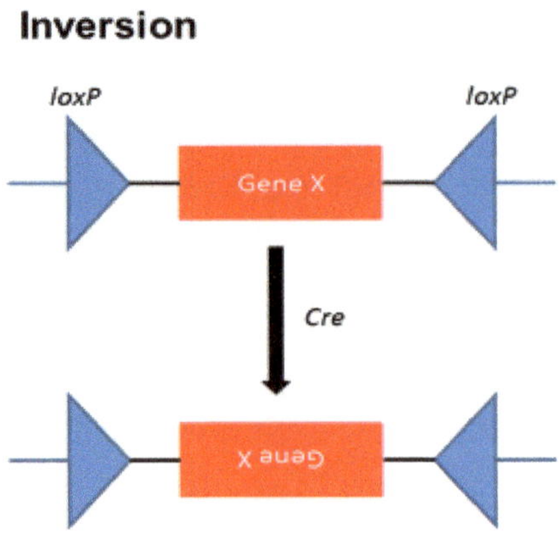

Cre-lox inversion:

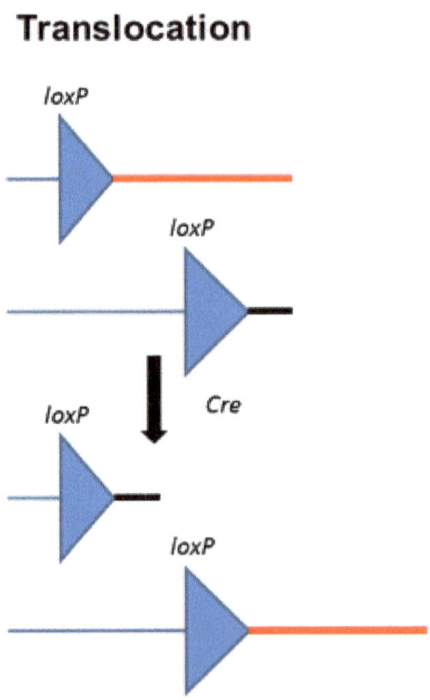

Cre-lox translocation:

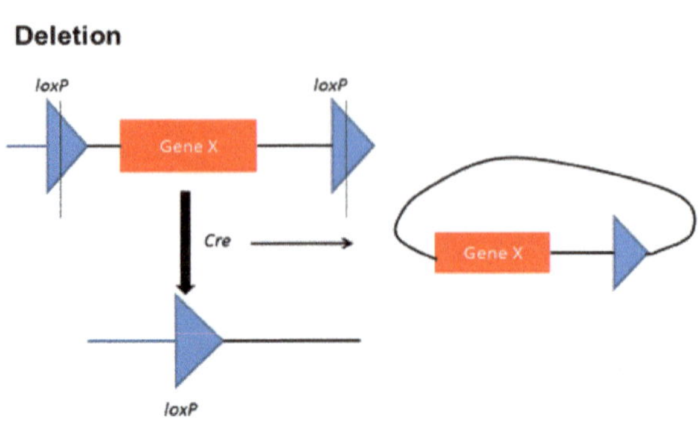

Cre-lox deletion:

Cre-Lox Recombination

Cre-Lox recombination is a site-specific recombinase technology, used to carry out deletions, insertions, translocations and inversions at specific sites in the DNA of cells. It allows the DNA modification to be targeted to a specific cell type or be triggered by a specific external stimulus. It is implemented both in eukaryotic and prokaryotic systems. The Cre-lox recombination system has been particularly useful to help neuroscientists to study the brain in which complex cell types and neural circuits come together to generate cognition and behaviors. NIH Blueprint for Neuroscience Research has created several hundreds of Cre driver mouse lines which are currently used by the world-wide neuroscience community.

The system consists of a single enzyme, Cre recombinase, that recombines a pair of short target sequences called the *Lox* sequences. This system can be implemented without inserting any extra supporting proteins or sequences. The Cre enzyme and the original *Lox* site called the *LoxP* sequence are derived from bacteriophage P1.

Placing Lox sequences appropriately allows genes to be activated, repressed, or exchanged for other genes. At a DNA level many types of manipulations can be carried out. The activity of the Cre enzyme can be controlled so that it is expressed in a particular cell type or triggered by an external stimulus like a chemical signal or a heat shock. These targeted DNA changes are useful in cell lineage tracing and when mutants are lethal if expressed globally.

The Cre-Lox system is very similar in action and in usage to the FLP-FRT recombination system.

History

Cre-Lox recombination is a special type of site-specific recombination developed by Dr. Brian Sauer initially for use in activating gene expression in mammalian cell lines (DuPont). Subsequently, researchers in the laboratory of Dr. Jamey Marth demonstrated that Cre-Lox recombination could be used to delete loxP-flanked chromosomal DNA sequences at high efficiency in specific developing T-cells of transgenic animals, with the authors proposing that this approach could be used to define endogenous gene function in specific cell types, indelibly mark progenitors in cell fate determination studies, induce specific chromosomal rearrangements for biological and disease modeling, and determine the roles of early genetic lesions in disease (and phenotype) maintenance.

Shortly thereafter, researchers in the laboratory of Dr. Klaus Rajewsky reported the production of pluripotent embryonic stem cells bearing a targeted loxP-flanked (floxed) DNA polymerase gene. Combining these advances in collaboration, the laboratories of Drs. Marth and Rajewsky reported in 1994 that Cre-lox recombination could be used for conditional gene targeting in developing T-cells in mice. However, they observed only ~50% of the DNA polymerase beta gene was deleted in T cells based on DNA blotting. It was unclear whether only one allele in each T-cell or 50% of T cells had 100% deletion in both alleles. Researchers have since reported more efficient Cre-Lox conditional gene mutagenesis in the developing T cells by the Marth laboratory in 1995.

Independently, Joe Z. Tsien has pioneered the use of Cre-loxP system for cell type- and region-specific gene manipulation in the adult brain where hundreds of distinct neuron types may exist and nearly all neurons in the adult brain are known to be post-mitotic. Tsien and his colleagues demonstrated Cre-mediated recombination can occur in the post-mitotic pyramidal neurons in the adult mouse forebrain.

These developments have led to a wide-spread use of conditional mutagenesis in biomedical research, spanning many disciplines in which it becomes a powerful platform for determining gene function in specific cell types and at specific developmental times. In particular, the clear demonstration of its usefulness in precisely defining the complex relationship between specific cells/circuits and behaviors for brain research, has promoted the NIH to initiate the NIH Blueprint for Neuroscience Research Cre-driver mouse projects in early 2000. To date, NIH Blueprint for Neuroscience Research Cre projects have created several hundreds of Cre driver mouse lines which are currently used by the world-wide neuroscience community.

Overview

Cre-Lox recombination involves the targeting of a specific sequence of DNA and splicing it with the help of an enzyme called Cre recombinase. Because systemic inactivation of many genes further causes embryonic lethality, Cre-Lox recombination is commonly used to circumvent this problem. In addition, Cre–Lox recombination provides the best experimental control that presently exists in transgenic animal modeling to link genotypes (alterations in genomic DNA) to phenotypes (the biological outcomes).

The Cre-lox system is used as a genetic tool to control site specific recombination events in genomic DNA. This system has allowed researchers to manipulate a variety of genetically modified organisms to control gene expression, delete undesired DNA sequences and modify chromosome architecture.

The Cre protein is a site-specific DNA recombinase, that can catalyse the recombination of DNA between specific sites in a DNA molecule. These sites, known as *loxP* sequences, contain specific binding sites for Cre that surround a directional core sequence where recombination can occur.

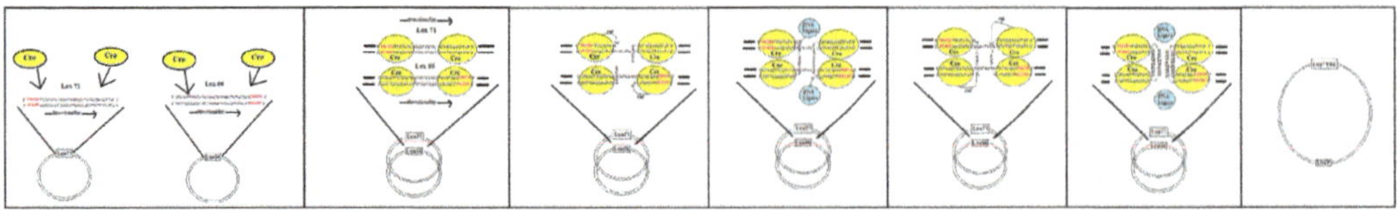

A diagram describing how Lox71 and Lox66 sites can be used to combine two plasmids into one contiguous plasmid.

When cells that have *loxP* sites in their genome express Cre, a recombination event can occur between the *loxP* sites. Cre recombinase proteins bind to the first and last 13 bp regions of a lox site forming a dimer. This dimer then binds to a dimer on another lox site to form a tetramer. Lox sites are directional and the two sites joined by the tetramer are parallel in orientation. The double stranded DNA is cut at both *loxP* sites by the Cre protein. The strands are then rejoined with DNA ligase in a quick and efficient process. The result of recombination depends on the orientation of the *loxP* sites. For two lox sites on the same chromosome arm, inverted *loxP* sites will cause an inversion of the intervening DNA, while a direct repeat of *loxP* sites will cause a deletion event. If *loxP* sites are on different chromosomes it is possible for translocation events to be catalysed by Cre induced recombination. Two plasmids can be joined using the variant lox sites 71 and 66.

Cre Recombinase

The Cre protein (encoded by the locus originally named as "Causes recombination", with "Cyclization recombinase" being found in some references) consists of 4 subunits and two domains: The larger carboxyl (C-terminal) domain, and smaller amino (N-terminal) domain. The total protein has 343 amino acids. The C domain is similar in structure to the domain in the Integrase family of enzymes isolated from lambda phage. This is also the catalytic site of the enzyme.

loxP Site

loxP (locus of X-over P1) is a site on the bacteriophage P1 consisting of 34 bp. The site includes an asymmetric 8 bp sequence, variable except for the middle two bases, in between two sets of palindromic, 13 bp sequences. The exact sequence is given below; 'N' indicates bases which may vary, and lowercase letters indicate bases that have been mutated from the wild-type.

13bp	8bp	13bp
ATAACTTCGTATA -	NNNTANNN	-TATACGAAGTTAT

Example Alternate loxP Sites			
Name	**13bp Recognition Region**	**8bp Spacer Region**	**13bp Recognition Region**
Wild-Type	ATAACTTCGTATA	ATGTATGC	TATACGAAGTTAT
lox 511	ATAACTTCGTATA	ATGTATaC	TATACGAAGTTAT
lox 5171	ATAACTTCGTATA	ATGTgTaC	TATACGAAGTTAT
lox 2272	ATAACTTCGTATA	AaGTATcC	TATACGAAGTTAT
M2	ATAACTTCGTATA	AgaaAcca	TATACGAAGTTAT
M3	ATAACTTCGTATA	taaTACCA	TATACGAAGTTAT
M7	ATAACTTCGTATA	AgaTAGAA	TATACGAAGTTAT

M11	ATAACTTCGTATA	aGATAgaa	TATACGAAGTTAT
lox 71	taccgTTCGTATA	NNNTANNN	TATACGAAGTTAT
lox 66	ATAACTTCGTATA	NNNTANNN	TATACGAAcggta

Holliday Junctions and Homologous Recombination

During genetic recombination, a Holliday junction is formed between the two strands of DNA and a double-stranded break in a DNA molecule leaves a 3'OH end exposed. This reaction is aided with the endonuclease activity of an enzyme. 5' Phosphate ends are usually the substrates for this reaction, thus extended 3' regions remain. This 3' OH group is highly unstable, and the strand on which it is present must find its complement. Since homologous recombination occurs after DNA replication, two strands of DNA are available, and thus, the 3' OH group must pair with its complement, and it does so, with an intact strand on the other duplex. Now, one point of crossover has occurred, which is what is called a Holliday Intermediate.

The 3'OH end is elongated (that is, bases are added) with the help of DNA Polymerase. The pairing of opposite strands is what constitutes the crossing-over or Recombination event, which is common to all living organisms, since the genetic material on one strand of one duplex has paired with one strand of another duplex, and has been elongated by DNA polymerase. Further cleavage of Holliday Intermediates results in formation of Hybrid DNA.

This further cleavage or 'resolvation' is done by a special group of enzymes called Resolvases. RuvC is just one of these Resolvases that have been isolated in bacteria and yeast.

For many years, it was thought that when the Holliday junction intermediate was formed, the branch point of the junction (where the strands cross over) would be located at the first cleavage site. Migration of the branch point to the second cleavage site would then somehow trigger the second half of the pathway. This model provided convenient explanation for the strict requirement for homology between recombining sites, since branch migration would stall at a mismatch and would not allow the second strand exchange to occur. In more recent years, however, this view has been challenged, and most of the current models for Int, Xer, and Flp recombination involve only limited branch migration 1–3 base pairs) of the Holliday intermediate, coupled to an isomerisation event that is responsible for switching the strand cleavage specificity.

Site-specific Recombination

Site-specific recombination (SSR) involves specific sites for the catalyzing action of special enzymes called recombinases. Cre, or cyclic recombinase, is one such enzyme. Site-specific recombination is, thus, the enzyme-mediated cleavage and ligation of two defined deoxynucleotide sequences.

A number of conserved site-specific recombination systems have been described in both prokaryotic and eukaryotic organisms. In general, these systems use one or more proteins and act on unique asymmetric DNA sequences. The products of the recombination event depend on the relative orientation of these asymmetric sequences. Many other proteins apart from the recombinase are involved in regulating the reaction. During site-specific DNA recombination, which brings about genetic rearrangement in processes such as viral integration and excision and chromosomal segregation, these recombinase enzymes recognize specific DNA sequences and catalyse the reciprocal exchange of DNA strands between these sites.

Mechanism of Action

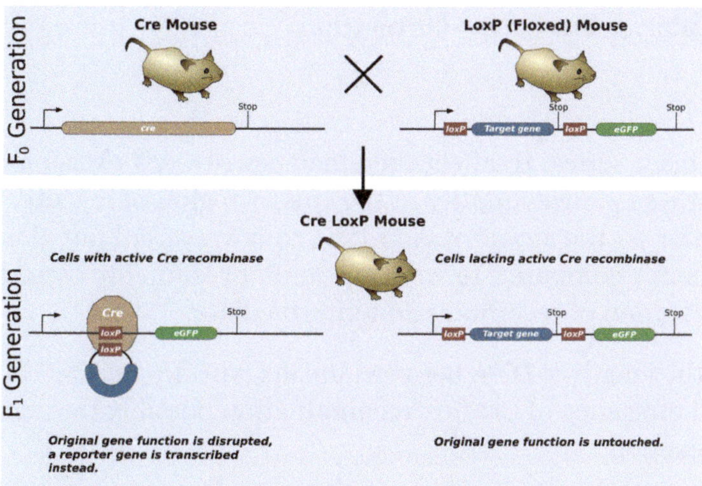

A model experiment in genetics using the Cre-lox system: the premature stop sequence present in floxed mice is removed only from cells that express Cre recombinase when the mice are bred together

Initiation of site-specific recombination begins with the binding of recombination proteins to their respective DNA targets. A separate recombinase recognizes and binds to each of two recombination sites on two different DNA molecules or within the same DNA strand. At the given specific site on the DNA, the hydroxyl group of the tyrosine in the recombinase attacks a phosphate group in the DNA backbone using a direct transesterification mechanism. This reaction links the recombinase protein to the DNA via a phospho-tyrosine linkage. This conserves the energy of the phosphodiester bond, allowing the reaction to be reversed without the involvement of a high-energy cofactor.

Cleavage on the other strand also causes a phospho-tyrosine bond between DNA and the enzyme. At both of the DNA duplexes, the bonding of the phosphate group to tyrosine residues leave a 3' OH group free in the DNA backbone. In fact, the enzyme-DNA complex is an intermediate stage, which is followed by the ligation of the 3' OH group of one DNA strand to the 5' phosphate group of the other DNA strand, which is covalently bonded to the tyrosine residue; that is, the covalent linkage between 5' end and tyrosine residue is broken. This reaction synthesizes the Holliday junction discussed earlier.

In this fashion, opposite DNA strands are joined together. Subsequent cleavage and rejoining cause DNA strands to exchange their segments. Protein-protein interactions drive and direct strand exchange. Energy is not compromised, since the protein-DNA linkage makes up for the loss of the phosphodiester bond, which occurred during cleavage.

Site-specific recombination is also an important process that viruses, such as bacteriophages, adopt to integrate their genetic material into the infected host. The virus, called a *prophage* in such a state, accomplishes this via integration and excision. The points where the integration and excision reactions occur are called the attachment (att) sites. An attP site on the phage exchanges segments with an attB site on the bacterial DNA. Thus, these are site-specific, occurring only at the respective att sites. The integrase class of enzymes catalyse this particular reaction.

Efficiency of Action

Two factors have been shown to affect the efficiency of Cre's excision on the lox pair. First, the nucleotide sequence identity in the spacer region of lox site. Engineered lox variants which differ on the spacer region tend to have varied but generally lower recombination efficiency compared to wildtype loxP, presumably through affecting the formation and resolution of recombination intermediate.

Another factor is the length of DNA between the lox pair. Increasing the length of DNA leads to decreased efficiency of Cre/lox recombination possibly through regulating the dynamics of the reaction.

Natural Function of the Cre-lox System

The P1 phage is a temperate phage that causes either a lysogenic or lytic cycle when it infects a bacterium. In its lytic state, once its viral genome is injected into the host cell, viral proteins are produced, virions are assembled, and the host cell is lysed to release the phages, continuing the cycle. In the lysogenic cycle the phage genome replicates with the rest of the bacterial genome and is transmitted to daughter cells at each subsequent cell division. It can transition to the lytic cycle by a later event such as UV radiation or starvation.

Phages like the lambda phage use their site specific recombinases to integrate their DNA into the host genome during lysogeny. P1 phage DNA on the other hand, exists as a plasmid in the host. The Cre-lox system serves several functions in the phage: it circularizes the phage DNA into a plasmid, separates interlinked plasmid rings so they are passed to both daughter bacteria equally and may help maintain copy numbers through an alternative means of replication.

The P1 phage DNA when released into the host from the virion is in the form of a linear double stranded DNA molecule. The Cre enzyme targets loxP sites at the ends of this molecule and cyclises the genome. This can also take place in the absence of the Cre lox system with the help of other bacterial and viral proteins. The P1 plasmid is relatively large (~90Kbp) and hence exists in a low copy number - usually one per cell. If the two daughter plasmids get interlinked one of the daughter cells of the host will lose the plasmid. The Cre-lox recombination system prevents these situations by unlinking the rings of DNA by carrying out two recombination events (linked rings -> single fused ring -> two unlinked rings). It is also proposed that rolling circle replication followed by recombination will allow the plasmid to increase its copy number when certain regulators (repA) are limiting.

Implementation of Multiple Loxp Site Pairs

Multiple variants of loxP, in particular lox2272 and loxN, have been used by researchers with the combination of different Cre actions (transient or constitutive) to create a "Brainbow" system that allows multi-colouring of mice's brain with four fluorescent proteins.

Another report using two lox variants pair but through regulating the length of DNA in one pair results in stochastic gene activation with regulated level of sparseness.

Humanized Mouse

A mouse carrying functional human genes or cells is called humanized. Humanized mice are made by introducing the functional copy of a human gene into a recipient mouse in the form of expression plasmid. The mice containing the functional copy of a human gene are bred with a knockout mouse. Alternatively the humanized mouse can also be made by directly injecting the human transgene into pronuclear embryos from mice that are knockout for the corresponding murine gene. In addition, the human gene can also be introduced using a routine homologous recombination in embryonic stem cells with the corresponding mouse gene.

Knockout Mouse

A knockout mouse or knock-out mouse is a genetically modified mouse (*Mus musculus*) in which researchers have inactivated, or "knocked out", an existing gene by replacing it or disrupting it with an artificial piece of DNA. They are important animal models for studying the role of genes which have been sequenced but whose functions have not been determined. By causing a specific gene to be inactive in the mouse, and observing any differences from normal behaviour or physiology, researchers can infer its probable function.

Mice are currently the laboratory animal species most closely related to humans for which the knockout technique can easily be applied. They are widely used in knockout experiments, especially those investigating genetic questions that relate to human physiology. Gene knockout in rats is much harder and has only been possible since 2003.

The first recorded knockout mouse was created by Mario R. Capecchi, Martin Evans, and Oliver Smithies in 1989, for which they were awarded the 2007 Nobel Prize in Physiology or Medicine. Aspects of the technology for generating knockout mice, and the mice themselves have been patented in many countries by private companies.

Use

Knocking out the activity of a gene provides information about what that gene normally does. Humans share many genes with mice. Consequently, observing the characteristics of knockout mice gives researchers information that can be used to better understand how a similar gene may cause or contribute to disease in humans.

Examples of research in which knockout mice have been useful include studying and modeling different kinds of cancer, obesity, heart disease, diabetes, arthritis, substance abuse, anxiety, aging and Parkinson's disease. Knockout mice also offer a biological and scientific context in which drugs and other therapies can be developed and tested.

Millions of knockout mice are used in experiments each year.

Strains

There are several thousand different strains of knockout mice. Many mouse models are named after the gene that has been inactivated. For example, the p53 knockout mouse is named after the p53 gene which codes for a protein that normally suppresses the growth of tumours by arresting cell division and/or inducing apoptosis. Humans born with mutations that deactivate the p53 gene suffer from Li-Fraumeni syndrome, a condition that dramatically increases the risk of developing bone cancers, breast cancer and blood cancers at an early age. Other mouse models are named according to their physical characteristics or behaviours.

Procedure

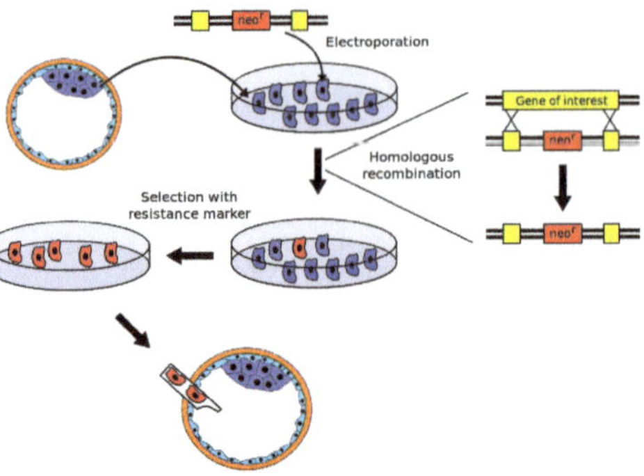

The procedure for making mixed-genotype blastocyst.

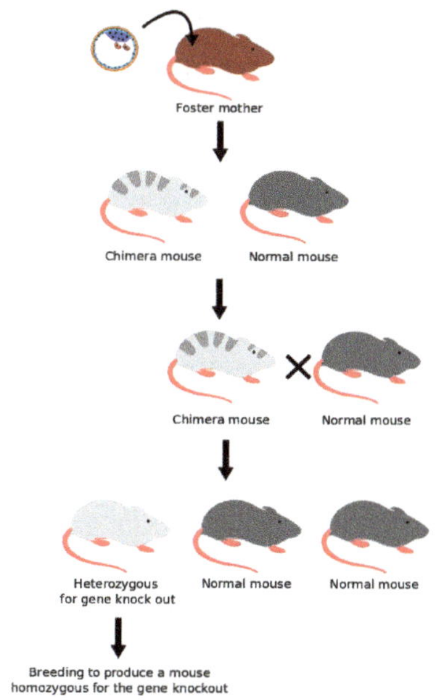

Breeding scheme for producing knockout mice. Blastocysts containing cells, that are both wildtype and knockout cells, are injected into the uterus of a foster mother. This produces offspring that are either wildtype and coloured the same colour as the blastocyst donor (grey) or chimera (mixed) and partially knocked out. The chimera mice are crossed with a normal wildtype mouse (grey). This produces offspring that are either white and heterozygous for the knocked out gene or grey and wildtype. White heterozygous mice can subsequently be crossed to produce mice that are homozygous for the knocked out gene.

There are several variations to the procedure of producing knockout mice; the following is a typical example.

1. The gene to be knocked out is isolated from a mouse gene library. Then a new DNA sequence is engineered which is very similar to the original gene and its immediate neighbour sequence, except that it is changed sufficiently to make the gene inoperable. Usually, the new sequence is also given a marker gene, a gene that normal mice don't have and that confers resistance to a certain toxic agent (e.g., neomycin) or that produces an observable change (e.g. colour or fluorescence). In addition, a second gene, such as herpes tk+, is also included in the construct in order to accomplish a complete selection.

2. Embryonic stem cells are isolated from a mouse blastocyst (a very young embryo) and grown *in vitro*. For this example, we will take stem cells from a white mouse.

3. The new sequence from step 1 is introduced into the stem cells from step 2 by electroporation. By the natural process of homologous recombination some of the electroporated stem cells will incorporate the new sequence with the knocked-out gene into their chromosomes in place of the original gene. The chances of a successful recombination event are relatively low, so the major-

ity of altered cells will have the new sequence in only one of the two relevant chromosomes – they are said to be heterozygous. Cells that were transformed with a vector containing the neomycin resistance gene and the herpes tk+ gene are grown in a solution containing neomycin and Ganciclovir in order to select for the transformations that occurred via homologous recombination. Any insertion of DNA that occurred via random insertion will die because they test positive for both the neomycin resistance gene and the herpes tk+ gene, whose gene product reacts with Ganciclovir to produce a deadly toxin. Moreover, cells that do not integrate any of the genetic material test negative for both genes and therefore die as a result of poisoning with neomycin.

4. The embryonic stem cells that incorporated the knocked-out gene are isolated from the unaltered cells using the marker gene from step 1. For example, the unaltered cells can be killed using a toxic agent to which the altered cells are resistant.

5. The knocked-out embryonic stem cells from step 4 are inserted into a mouse blastocyst. For this example, we use blastocysts from a grey mouse. The blastocysts now contain two types of stem cells: the original ones (from the grey mouse), and the knocked-out cells (from the white mouse). These blastocysts are then implanted into the uterus of female mice, where they develop. The newborn mice will therefore be chimeras: some parts of their bodies result from the original stem cells, other parts from the knocked-out stem cells. Their fur will show patches of white and grey, with white patches derived from the knocked-out stem cells and grey patches from the recipient blastocyst.

6. Some of the newborn chimera mice will have gonads derived from knocked-out stem cells, and will therefore produce eggs or sperm containing the knocked-out gene. When these chimera mice are crossbred with others of the wild type, some of their offspring will have one copy of the knocked-out gene in all their cells. These mice will be entirely white and are not chimeras, however they are still heterozygous.

7. When these heterozygous offspring are interbred, some of their offspring will inherit the knocked-out gene from both parents; they carry no functional copy of the original unaltered gene (i.e. they are homozygous for that allele).

A detailed explanation of how knockout (KO) mice are created is located at the website of the Nobel Prize in Physiology or Medicine 2007.

Limitations

The National Institutes of Health discusses some important limitations of this technique.

While knockout mouse technology represents a valuable research tool, some important limitations exist. About 15 percent of gene knockouts are developmentally lethal, which means that the genetically altered embryos cannot grow into adult mice. This problem is often overcome through the use of conditional mutations. The lack of adult mice limits studies to embryonic development and often makes it more difficult to determine a gene's function in relation to human health. In some instances, the gene may serve a different function in adults than in developing embryos.

Knocking out a gene also may fail to produce an observable change in a mouse or may even produce different characteristics from those observed in humans in which the same gene is inactivated. For example, mutations in the p53 gene are associated with more than half of human cancers and often lead to tumours in a particular set of tissues. However, when the p53 gene is knocked out in mice, the animals develop tumours in a different array of tissues.

There is variability in the whole procedure depending largely on the strain from which the stem cells have been derived. Generally cells derived from strain 129 are used. This specific strain is not suitable for many experiments (e.g., behavioural), so it is very common to backcross the offspring to other strains. Some genomic loci have been proven very difficult to knock out. Reasons might be the presence of repetitive sequences, extensive DNA methylation, or heterochromatin. The confounding presence of neighbouring 129 genes on the knockout segment of genetic material has been dubbed the "flanking-gene effect". Methods and guidelines to deal with this problem have been proposed.

Another limitation is that conventional (i.e. non-conditional) knockout mice develop in the absence of the gene being investigated. At times, loss of activity during development may mask the role of the gene in the adult state, especially if the gene is involved in numerous processes spanning development. Conditional/inducible mutation approaches are then required that first allow the mouse to develop and mature normally prior to ablation of the gene of interest.

Another serious limitation is a lack of evolutive adaptations in knockout model that might occur in wild type animals after they naturally mutate. For instance, erythrocyte-specific coexpression of GLUT1 with stomatin constitutes a compensatory mechanism in mammals that are unable to synthesize vitamin C.

References

- Urbina, Zachary (12 February 2013) "Genetically Engineered Virus Fights Liver Cancer" United Academics, Retrieved 15 February 2013
- Sanford, JC; et al. (1987). "Delivery of substances into cells and tissues using a particle bombardment process". Journal of Particulate Science and Technology. 5: 27–37. doi:10.1080/02726358708904533

- "Treatment for Leukemia Is Showing Early Promise". The New York Times. Associated Press. 11 August 2011. p. A15. Retrieved 21 January 2013
- Panesar, Pamit et al. (2010) Enzymes in Food Processing: Fundamentals and Potential Applications, Chapter 10, I K International Publishing House, ISBN 978-93-80026-33-6
- Pollack, Andrew (7 May 2014). "Researchers Report Breakthrough in Creating Artificial Genetic Code". New York Times. Retrieved 7 May 2014
- Gasdaska JR et al. (2003) "Advantages of Therapeutic Protein Production in the Aquatic Plant Lemna". BioProcessing Journal Mar/Apr 2003 pp 49–56
- Knapton, Sarah (6 February 2016). "Releasing millions of GM mosquitoes 'could solve zika crisis'". The Telegraph. Retrieved 14 March 2016
- Rieger R. Michaelis A., Green M. M. (1976). Glossary of genetics and cytogenetics: Classical and molecular. Heidelberg - New York: Springer-Verlag. ISBN 3-540-07668-9
- Nicholls, Henry (14 September 2011) "Swarm troopers: Mutant armies waging war in the wild" The New Scientist. Retrieved 20 September 2011
- "'Any idiot can do it.' Genome editor CRISPR could put mutant mice in everyone's reach". Science | AAAS. 2016-11-02. Retrieved 2016-12-02
- Classical Medicine Journal (14 April 2010). "Genetically modified cows producing human milk.". Archived from the original on 6 November 2014
- King R. C., Stransfield W. D. (1998): Dictionary of genetics. Oxford University Press, New York, Oxford, ISBN 0-19-50944-1-7; ISBN 0-19-509442-5
- Whipple, Tom (2016-05-14). "Trailblazing GM vaccines 'are held back by red tape'". www.thetimes.co.uk. The Times. Retrieved 2016-05-14
- Summers, Rebecca (24 April 2013) "Bacteria churn out first ever petrol-like biofuel" New Scientist, Retrieved 27 April 2013
- M. K. Sateesh (25 August 2008). Bioethics And Biosafety. I. K. International Pvt Ltd. pp. 456–. ISBN 978-81-906757-0-3. Retrieved 27 March 2013
- JoAnna Wendel for the Genetic Literacy Project. 10 September 2013 Scientists, journalists and farmers join lively GMO forum
- Molteni, Megan (2017-04-12). "Florida's Orange Trees Are Dying, But a Weaponized Virus Could Save Them". WIRED. Retrieved 2017-04-17

Permissions

All chapters in this book are published with permission under the Creative Commons Attribution Share Alike License or equivalent. Every chapter published in this book has been scrutinized by our experts. Their significance has been extensively debated. The topics covered herein carry significant information for a comprehensive understanding. They may even be implemented as practical applications or may be referred to as a beginning point for further studies.

We would like to thank the editorial team for lending their expertise to make the book truly unique. They have played a crucial role in the development of this book. Without their invaluable contributions this book wouldn't have been possible. They have made vital efforts to compile up to date information on the varied aspects of this subject to make this book a valuable addition to the collection of many professionals and students.

This book was conceptualized with the vision of imparting up-to-date and integrated information in this field. To ensure the same, a matchless editorial board was set up. Every individual on the board went through rigorous rounds of assessment to prove their worth. After which they invested a large part of their time researching and compiling the most relevant data for our readers.

The editorial board has been involved in producing this book since its inception. They have spent rigorous hours researching and exploring the diverse topics which have resulted in the successful publishing of this book. They have passed on their knowledge of decades through this book. To expedite this challenging task, the publisher supported the team at every step. A small team of assistant editors was also appointed to further simplify the editing procedure and attain best results for the readers.

Apart from the editorial board, the designing team has also invested a significant amount of their time in understanding the subject and creating the most relevant covers. They scrutinized every image to scout for the most suitable representation of the subject and create an appropriate cover for the book.

The publishing team has been an ardent support to the editorial, designing and production team. Their endless efforts to recruit the best for this project, has resulted in the accomplishment of this book. They are a veteran in the field of academics and their pool of knowledge is as vast as their experience in printing. Their expertise and guidance has proved useful at every step. Their uncompromising quality standards have made this book an exceptional effort. Their encouragement from time to time has been an inspiration for everyone.

The publisher and the editorial board hope that this book will prove to be a valuable piece of knowledge for students, practitioners and scholars across the globe.

Index